Given in memory of
Patricia T. Alford
by
JoAnn and Juluis Fouts
Williamsburg
Regional Library

Passionate Minds

Passionate Minds

Women Rewriting the World

Claudia Roth Pierpont

Alfred A. Knopf New York 2000

This Is a Borzoi Book
Published by Alfred A. Knopf

 Published in the United States by Alfred A. Knopf, a division of Random House, Inc., New York, and simultaneously in Canada by Random House of Canada Limited, Toronto. Distributed by Random House, Inc., New York.

www.randomhouse.com

All the essays in this book were originally published, in somewhat different form, in *The New Yorker.*

Knopf, Borzoi Books, and the colophon are registered trademarks of Random House, Inc.

Library of Congress Cataloging-in-Publication Data
Pierpont, Claudia Roth.
Passionate minds : women rewriting the world / by Claudia Roth Pierpont. — 1st ed.
p. cm.
Includes index.
ISBN 0-679-43106-3
1. American literature—Women authors—History and criticism. 2. Women and literature—United States—History—20th century. 3. Women and literature—Great Britain—History—20th century. 4. English literature—Women authors—History and criticism. 5. American literature—20th century—History and criticism. 6. English literature—20th century—History and criticism. 7. Women authors, American—20th century Biography. 8. Women authors, English—20th century Biography. I. Title.
PS 151 P54 2000
810.9'9287'0904—dc21 99-33349
CIP

Manufactured in the United States of America
First Edition

To Julia and to Shirley

Contents

Acknowledgments

This book would not exist without Robert Gottlieb, who brought me to *The New Yorker,* and whose memorable instructions on how to deal with the exciting and complicated issues that seemed to spring in all directions from a simple book-review assignment were: "Give our readers Olive Schreiner. Make it as long as you need to." His brilliance and enthusiasm as an editor are rightly famous, and I would like to add my particular gratitude for the spirit of adventure that allowed him to entrust a fairly untried writer with such unbounded opportunities, and the conspiratorial delight with which he greeted results at every turn. He has been more than an editor; he has been an education.

My gratitude to Alice Truax, at *The New Yorker,* is also immense. Her intelligence and concentration, her insight and dedication, have been a guide, a corrective, and an enormous pleasure to work with for more than five years.

All of the essays here have appeared in some form in *The New Yorker;* several have received additions and revisions for this book, while two—on Doris Lessing, and on Hannah Arendt and Mary McCarthy—have been radically expanded. I am grateful to Tina Brown and to David Remnick for their support, and indebted to many extraordinary people who work or have worked at the magazine, and who have provided everything from astute overall criticism to a better choice of word to grammatical repair to a correction of vital fact to a way of climbing out of a bleak and hopeless literary pit: to Henry Finder, Ann Goldstein, Eleanor Gould Packard, Charles McGrath, Louis Menand, Elizabeth Macklin, Dorothy Wickenden, Deborah Garrison, Elizabeth Pearson-Griffiths, Jim Albrecht, Martin Baron, and Amy Tübke-Davidson, my sincere thanks.

I would also like to express my gratitude to the Whiting Foundation, and particularly to the memory of Gerald Freund, who cared so much about enabling writers to work.

My gratitude, too, for the gracious assistance of Lotte Köhler of the Hannah Arendt Literary Trust, and of Mary Adelman of venerable Osner Business Machines.

To Dr. Gregory K. Harmon, exemplary physician, gratitude beyond words for gifts of sight.

Thanks to irreplaceable booksellers and first-class talkers: Robert and Dorothy Emerson, Barbara Farnsworth, Michael P. McClure and Stephen M. Tebid.

A special loving thanks to Arlene Croce, who has opened up so many possibilities and been such a generous friend. Heartfelt thanks also to Scott Griffin and to Bonnie Yochelson, for listening and questioning and for making the process of writing much less lonely.

My thanks to my wonderful daughter, Julia Pierpont, for providing the subtitle of this book. And to her grandparents, Bob and Mary Pierpont, for all they have done.

Lastly, my deep and loving gratitude to Robert Pierpont, longtime first reader, for years of unflagging encouragement, and to Shirley Roth for a sometimes frightening wisdom.

Introduction

This book was conceived when two writers who seemed to embody entirely different concerns—the South African "agnostic" novelist Olive Schreiner and the author of America's champion best-seller, Margaret Mitchell—turned out to have a great deal in common. Born nearly half a century and half a world apart, both wrote about race and its place in the history of a bitterly divided country, about a time of difficult change from one era to the next, and about women who were too strong to fit into established feminine patterns. And both had an extraordinary effect on readers: in the 1880s, Schreiner's *The Story of an African Farm* was widely perceived as a spiritual bridge between Darwinian science and the need for God; in the mid-1930s and for a long time after, a significant number of Americans viewed the Civil War and Reconstruction—with all their implications for the contemporary racial order—as Mitchell had portrayed them in *Gone with the Wind.* Whatever the merits of their prose or their arguments, these women told stories that changed the way people thought and lived.

Fascinated by this kind of naïve power—both Schreiner's and Mitchell's books were first novels—I began to consider other literary women of influence (very different from women of literary influence) whose domain is somewhat off the usual critical path. The resulting group is emphatically diverse; there is hardly a woman here who would not be scandalized to find herself in company with most of the others. Hannah Arendt and Ayn Rand, Gertrude Stein and Mae West, Doris Lessing and Anaïs Nin, Zora Neale Hurston and Eudora Welty, Marina Tsvetaeva and Mary McCarthy: what could they possibly have in common? Part of the excitement of working with such contrary figures lay in comparing their different versions of absolute truth: the individualism of Rand versus the Communism of Lessing, the ideal of sexual liberation in Mae West versus that of Nin or Lessing or McCarthy, the voice of the American South as heard by Hurston and by Welty. And yet, again, similarities began to emerge, not in what these women wrote but in how they contrived to get it written: that is, in how ambitious women worked out their destinies in an age of momentous transition for their sex, when—to paraphrase Olive Schreiner on religious faith—the old ways seemed outworn but new ones had not been invented.

This book is organized into three sections within an overall (if elastic) chronological sequence. Broadly speaking, the first section deals with issues of sexual freedom (Schreiner, Stein, Nin, West), the second with race (Mitchell, Hurston, Welty), and the third with politics—particularly with the idea and reality of Communism (Tsvetaeva, Rand, Lessing, Arendt and McCarthy). There are, however, many overlapping interests and crossings of category. Marina Tsvetaeva, for example, might easily have been placed within the first group—for her life's lesson about sexual desire as a driving source of creativity—but her experience under Stalin offers a valuable check against some of the more theoretical responses to the Soviet state that follow. Mae West may not seem to belong in this book at all, but it is precisely the point that she was, by necessity, a writer as well as a performer; if she hadn't made herself up, characters and plays and scripts and jokes and all, no one would have done it for her. And this process of self-creation is only slightly less true—less obviously, literally true—for many of the others. (Can there be two grander, braver, more perversely endearing heroines in twentieth-century fiction than Gertrude Stein and Ayn Rand?)

Certain bases of comparison, then, are obvious: this book is constructed around them. Others appeared only as I went along, and these themes will emerge from story to story: the undermining dangers of romantic love, knowingly avoided or helplessly pursued; the vital, unused energies of an earlier generation of mothers who were strong enough to push (or weak enough to drive) their daughters out into the world; physical beauty as a useful weapon and as a mirror-lined trap; the question of a feminine style in writing, aspired to or despised; the opposition between moral and artistic purpose—or, as it was often defined, between proper womanly sacrifice and disgracefully unfemale selfishness. There were so many possibilities for error and catastrophe, all so eagerly embraced, and all likely to prompt furious exasperation from today's reader—along with the affectionate gratitude that these women deserve. For they had hardly any models to follow, apart from a handful of suicidal literary heroines; one of the reasons we can judge so harshly now is because we have had *them.*

It might reasonably be asked whether this book represents a progress. What are the advances made in the lives and attitudes of women over the course of a century, from Olive Schreiner to Doris Lessing? The rights and advantages that Schreiner fought for—the votes, the jobs—have been well established. But what of internal change? That has turned out to be the most difficult subject to grapple with, impossible to summarize because so distinct from life to life. These are, after all, women of a transitional age—as Schreiner defined it, as Lessing also experienced it, as many women still do. These are lives in which success is hard won, retreat and even breakdown are common, love is difficult, and children are nearly impossible, lives in which all that is ever certain is that books and plays and poems are being written.

Passionate Minds

A Woman's Place

Olive Schreiner

On a summer day in 1897, Canon Ernest Roland Wilberforce, a member of the great ecclesiastical family that had led the century's battles against slavery and against Darwinism, paid a call on a middle-aged South African lady visitor to London to ask her aid in bringing about the spiritual regeneration of England. The canon's confidence in the abilities of this lady, Olive Schreiner, was based on her novel *The Story of an African Farm,* whose appearance fourteen years earlier had been met with denunciations of blasphemy from some of the sterner Church of England pulpits and with almost compassionate accounts of the snares of agnosticism from more ecumenical platforms. By her own report, Schreiner had hardly set foot in a church since adolescence. In those progressive London quarters where the displaced energies of orthodoxy were being reinvested in the many compensatory religions of late Victorianism—art, philanthropy, socialism, the various ideals of human progress toward perfection on earth—Schreiner's book earned her the position of licensed freethinker, of a dispenser of Truth during the last historical moment when that substance was still widely viewed as sufficiently objective and monolithic to merit a capital initial.

But the canon—urged on, perhaps, by more than a decade's plummeting Sunday attendance—had his reasons. The continuing popularity of Schreiner's book, in edition after edition, was due largely to a constituency of the spiritual-minded unchurched, an educated and ever-growing population of reluctant unbelievers who felt themselves cast from the Heaven of their parents and grandparents by the revelations of science. For many of those to whom the irreconcilability of geology and Genesis amounted to a sorrowful enlightenment, an edict of liberation accompanied by a theft of purpose, Schreiner's novel became a kind of guidebook. Selling some hundred thousand copies between its publication in 1883 and the end of the century, *African Farm* was one of those rare books that are not merely read but copied out and shared and appealed to, a book that was spoken of as affecting courses of action, as providing certain of its readers with freedom of thought while depriving others of peace of mind. Written for an age that fell "between faith and reason," as one of her characters describes it, *African Farm* tempered an insistent rejection of illusion with a passion for the moral and emotional impulses of Christianity. Its message was complex and forceful, and vague enough to offer courage in unexpected directions: John Galsworthy dispensed it to friends troubled with religious doubt; Gladstone thought it proved that God is Love. For the majority of its readers, however, *African Farm* served as a means of internal fortification: it reduced and assuaged the powerful desire to kneel.

The religious theme in fiction was itself anything but new. Schreiner's book appeared, in fact, toward the end of "an age of Religious Novels"—so christened as early as 1846 by the *Dublin Review*, in tallying the new genre at more than a third of the previous year's output of fiction. "Theological romances" were written in demonstration of every conceivable point of contemporary faith and prejudice: anti-Tractarian and evangelical, antievangelical and Jesuitical, anti-Jesuitical (plenty of these) and Broad Church, High Church, and no church whatever, with even the last subjected to argumentative and hairsplitting distinctions among freethinkers, agnostics, and atheists. Whether set in the decadent imperial Rome of the early martyrs or in a modern-day England where conversions and lapses undid marriages and tore families apart, the dramatized doctrines and the bickering over incompatible salvations were attended to raptly by several generations, as a continuing celestial soap opera, until the inspirational wellsprings suddenly emptied and dried as the century drew to a close.

From this library gone wholly to quaintness and dust Schreiner's book stands out, in part because its position is more inquiring than dogmatic and its questions are so simple and urgent. Her principal characters, a saintly shepherd boy and a rebellious orphan girl, are consumed each by a single idea, forming the book's double theme—he by the injustice of God to men, she by the injustice of men to women. As characters, these two consist of little else, yet their story does not disintegrate into the dry set-piece polemics of Charles Kingsley or of Mrs. Humphry Ward. Olive Schreiner was a missionary's daughter, writing compulsively of her own laborious struggle for intellectual liberty. She started on *African Farm* when she was not much more than twenty years old and was still living in the orbit of her family and her early moral education, still virtually trembling with the effort of resistance and the agitation of discovery. Her achievement was to take the genre's formulaic admixture of story and argument (a contemporary reviewer listed the issues Schreiner raised as "Orthodox Christianity, Unitarian Christianity, woman suffrage, marriage, Malthusianism, and immortality") and reforge them in the heat of youth, under relentless emotional pressure. The book retains an awkward, self-conscious vitality, as the record of a difficult transitional passage not only in the nineteenth-century history of ideas but in the continual reawakening of those ideas, independently and personally, in maturing minds.

In the 1880s, it already seemed ironic and a little shameful that Schreiner's novel was published with the author's name given on the title page as "Ralph Iron." The ploy of using a male pseudonym in order to gain serious attention went back to the Brontës, but by the time *African Farm* appeared, three years after the death of George Eliot, the virile mask amounted to little more than a convention. Eliot herself—who as Mary Ann Evans had offered her own tart opinions of "silly novels by lady novelists"—had been identified after the sensation of her first books, and it was she who made possible the notion of the authoress as social sage; her later works, especially, were intended and accepted as examples of moral philosophy. As happened with the books of her predecessors, Schreiner's *African Farm* was scrutinized for signs of Ralph Iron's true identity—sexual and social—and the male disguise was soon undone. There was hardly a reviewer, in fact, who failed to note that the book was evidently the work of a woman. The deduction was not based on points of inflection and detail, as had happened with, say, Charlotte Brontë, who was given away by such minutiae as Grace Poole's attention to curtain rings (which went to prove that *Jane Eyre,* in the words of one sleuth, "could have been written only by a woman or an upholsterer"). With Schreiner the evidence was far more broad-based, inhering in the desperate earnestness with which the novel struck its second dominating theme: the crushing suppression of possibility in women's lives.

The subject of woman's place was only a little newer to fiction than that of God's absence, and they were not unrelated. For Schreiner's generation, the two had been set side by side in Eliot's novels. Yet Eliot's great works were concerned more with women's obligations than with their rights. If there was no God, it was all the more imperative to behave as though there were, in order to assure motives for ethical behavior and to stave off Germanic nihilism. Social proprieties were the new household gods—if households were to continue, there must be gods—and women were the keepers of the altar. According to Eliot, woman's greatest gift was for "a sublimer resignation" to her lot.

Olive Schreiner wrote *African Farm* at roughly the age at which George Eliot first refused to accompany her father to church. Duty had been the daily bread of Schreiner's missionary upbringing, resignation what she was meant to feel on seeing her brothers sent off to school. To Schreiner these were not noble abstractions but hard and unreasoning—and certainly unsublime—constrictions. If the gods were false, the keepers of the altars were not holy vestals but slaves or fools. Schreiner, who had little ability for characterization, spoke directly through her heroine, and, whatever the artistic failings of the method, the result for the reader was a startling access of self-consciousness, a woman's voice newly direct and insolent and bitter:

> The world tells us what we are to be, and shapes us by the ends it sets before us. To you it says—*Work!* and to us it says—*Seem!* . . . The less a woman has in her head, the lighter she is for climbing.
>
> Yes, we have power; and since we are not to expend it in tunnelling mountains, nor healing diseases, nor making laws, nor money, nor on any extraneous object, we expend it on *you.* You are our goods, our merchandise, our material for operating on; we buy you, we sell you, we make fools of you. . . . We are not to study law, nor science, nor art; so we study you.

Schreiner read Eliot's *The Mill on the Floss* when she was very young, and again after completing *African Farm;* she deemed it suffocating. As a woman who suffered chronically and excruciatingly from asthma, Schreiner was something of a specialist in suffocation. Canon Wilberforce may well have imagined that he would find in her another St. Teresa, capable of shaping the spiritual reform the age demanded (Teresa, after all, had begun as a novelist). But Wilberforce arrived on Schreiner's doorstep in an age which—as George Eliot had warned—could neither create nor support another Teresa, and an age in which even that true Teresa of "far-resonant action" was to be appropriated as patron saint of the epidemic of hysterical illness that disabled and devoured generations of women whose gifts were not for resignation. It was in reference to St. Teresa's career of mysterious physical afflictions that the French neurologist Charcot and, later, Breuer and Freud identified as "stigmata" their female patients' inexplicable ailments: convulsions, obstructed vision, amnesia, depression, loss of speech, anorexia (recently "discovered"), paralysis. Olive Schreiner's asthma was but the central woe in an extensive array of debilities that, as much as any other factor, determined the course of her life. Mental endowment was not excluded by hysterical disease, as Breuer was eager to explain; intelligence was, in fact, rather likely to exceed the average. What was rendered nearly impossible was achievement.

The title *The Story of an African Farm* is misleading insofar as these words suggest a chronicle of activity and livelihood and social community, of the prospering interaction of man and land. Schreiner's geography was as alien to English expectations as the remotest reaches of empire would allow: no lowing in the pastures, no rosy dairymaids up to their elbows in curds and whey, but a measureless expanse of arid land burning under a blank sky. Her African terrain was equally far from the beguilingly exotic turquoise-and-gold of the leopard hunt. That sort of adventure was best written in Piccadilly, as Schreiner explained; the artist in Africa "must squeeze the colour from his brush, and dip it into the grey pigments around him." Schreiner's book was the first to place the landscape of the vast and barren South African karroo—like Emily Brontë's moors or Willa Cather's Southwestern mesas—on the map of literature.

In *The Story of an African Farm* the desert itself collaborates in human despair, and it was not beyond English reviewers of the it-can't-happen-here variety to lay blame for the author's desolate vision on the broad and flat-topped African mountains, peakless and unaspiring, which directed thoughts not upward but outward in futile dispersal. Africa was the pathetic fallacy reversed: the dry wilderness imposing its own bleak expressive life on the vision of those who beheld it. In claiming this wasteland for literature, Schreiner provided the twentieth century with a landscape it would require:

> Week after week, month after month, the sun looked down from the cloudless sky, till the karroo-bushes were leafless sticks, broken into the earth, and the earth itself was naked and bare; and only the milk-bushes, like old hags, pointed their shrivelled fingers heavenwards, praying for the rain that never came.

Beneath this pitiless sky lie the scattered buildings of an ostrich farm belonging to the fat and irritable Tant' Sannie, a Boer woman whose occasional moderation of temper is based on the religious conviction that she is spied on and appraised through "chinks in the world above." Growing up beside Sannie's placid and freckled stepdaughter, Em, is Em's orphaned cousin, a girl named Lyndall, whose perfections of diminutive beauty, particularly in the matter of hands and feet, are offered for the reader's constant admiration and for piquant contrast with the girl's intellectual grit. (Olive Schreiner stood barely five feet tall; to the end, her husband, whose published account of his wife tends toward vituperative damnation, offered praise for her tiny hands and feet.) Lyndall yearns to go to school but is thwarted by her aunt.

In the farm outbuildings live a genuinely devout German overseer and his son, Waldo, the shepherd with his mind fixed on God. Waldo builds an altar out of stones and watches for a sign as his offering—the mutton chop of his dinner—melts into pools of fat under the only burning eye his Heaven reveals. Waldo is tormented by doubt—on matters ranging from contradictions in Scripture to the existence of evil—but he is a compulsive believer, a genuine religious spirit.

When Waldo's kindhearted father dies, the one loving parent these children have known is replaced by a near monster called Bonaparte Blenkins, a red-nosed swindler who flatters his way into Sannie's dumb affections. For an invented trespass Waldo is taken to the fuel house and horribly whipped, with Blenkins's assurances that he is acting as a father and in accord with God's will: "I hope the Lord will bless and sanctify to you what I am going to do to you." That night, alone and bleeding in the dark—"he had not dared to stand still, and had not known it would end"—Waldo prays, and his prayers meet, again, with silence.

At this turning point, Schreiner's narrative halts. In the novel's second section, we are returned to Waldo's spiritual search, taken up again from the beginning—this time as it unfolds in the mind of the boy himself.

Never an easy or vivid storyteller, Schreiner slips gratefully into the strangely modern device of a soft, crooning monologue of drifting consciousness, strewn with Waldo's childhood memories—"We pull the green leaves softly into pieces to see the silk threads run across"—and culminating in the night that he howls and prays in the fuel house, and reaches his conclusion: "Now we have no God."

Three years go by. A stranger rides onto the farm; he is the authorial voice on horseback, and the story he tells Waldo forms what Schreiner (and many of her readers) saw as the allegorical core of the book. In this story, a reinvented *Pilgrim's Progress,* a hunter searches for "a vast white bird," which is Truth, and which he has seen only as a reflection on the water. His search brings him through the Land of Absolute Negation and Denial; it requires the release of the seductive black bird called Immortality. Pressing onward, the hunter passes his life hacking a series of stairs out of the sheer, towering rock of a mountainside. He never reaches the top—never glimpses Truth—but dies in the happiness of knowing that others will climb higher, using the steps he has cut. At the last, a white feather drifts down into his hand.

Waldo, for one, is mightily impressed. The modern reader is in something of the position of the narrator observing the boy's overexcited reaction: "There was something superbly ridiculous, unless one chanced to view it in another light." The philosophical Stranger leaves the boy with one last lesson: "To all who have been born in the old faith there comes a time of danger, when the old slips from us, and we have not yet planted our feet on the new." But that new life will come someday, although "we of this generation are not destined to eat and be satisfied as our fathers were; we must be content to go hungry." Then the Stranger pulls a book from his saddlebag—"It may give you a centre round which to hang your ideas"—and rides off.

In the years that follow, the proud and independent Lyndall fulfills her dream by getting herself sent away to school, but she returns disillusioned and bitter. Girls' finishing schools are machines for compressing the human soul into the smallest possible area: "I have seen some souls so compressed that they would have fitted into a small thimble." She wears a ring set with diamonds but claims that it betokens no engagement: "I am not in so great a hurry to put my neck beneath any man's foot; and I do not so greatly admire the crying of babies." Her sardonic awareness of her beauty almost redeems the author's Thumbelina descriptions: Lyndall allows herself the pleasure of thought only "when I am not too busy to find a new way of doing my hair that will show my little neck to better advantage," as she says. "Professional duties always first, you know."

Finally, Waldo goes off to see the world. Lyndall's ring-giver appears, and though she refuses to marry him—she does not love him—she agrees to run away with him, deep into the Transvaal. Lyndall, one finally learns, is pregnant. Her baby dies after three hours, and—in a singular touch of Victorian madness—Em's beau, who has fallen in love with Lyndall, dresses in women's clothing in order to be near her and to nurse her through her last illness. Lyndall, aged seventeen, dies rather vaguely but presumably of complications from childbirth. Her final epigrammatic insights (Gladstone copied them out in his diary) show a new, resigned grace: "Holiness is an infinite compassion for others," and the seemingly most un-Lyndall-like "Happiness is a great love and much serving."

When Waldo returns and hears the news, all the religious consolations he has known or read of race through his head, concluding with:

> The soul's fierce cry for immortality is this,—only this:—Return to me after death the thing as it was before. . . . Your immortality is annihilation, your Hereafter is a lie. . . . All dies, all dies! the roses are red with the matter that once reddened the cheek of the child.

Only in the obliterating unity of the cycles of nature, this flux where death is change, is there any comfort. Almost at peace with the thought, Waldo goes out to sit in the sunshine at the side of the wagon house among the startled, scattering chickens. And he dies there, just sitting, and of no apparent cause. Schreiner doesn't even propose a broken heart. The death is very gently managed: lifted, actually, from the scene in *Wuthering Heights* where Nelly Dean goes to tell Heathcliff of Cathy's death—he, of course, already knows—and finds him having stood all night so still against the old ash tree that the dew has soaked his hair and the birds pass around him fearlessly. The homely chickens of the African farm—now Em's farm—are so sure that this particular being poses no further threat that they clamber freely onto his shoulders and hands and nestle to sleep in his coat sleeves. Em, who comes to set a glass of milk by his side, says to herself, "He will wake soon." Olive Schreiner concludes, "But the chickens were wiser."

Lyndall—a name that *African Farm* made popular for South African girls—was the family name of Schreiner's mother, Rebekah Lyndall. The daughter of a London evangelical minister, Rebekah attributed her missionary commitment to the fervor aroused during an antiphonal hymn at a revival meeting: "Who will go and join the throng?" "*I* will go and join the throng." She was, by all accounts, well educated and intelligent, an accomplished young woman. In 1837 she married Gottlob Schreiner, a Lutheran from Germany, a former shoemaker who had been sent to London to continue his missionary training. Rebekah was nineteen. When she and her new husband entered the vestry to sign their names, the minister tore the wreath off her bonnet as unbefitting a missionary's bride. The couple set off for Africa three weeks later—just a few months after Victoria came to the throne and three years before Dr. Livingstone embarked for the "Dark Continent."

In Africa, Rebekah bore twelve children in twenty-four years. Five died in infancy or very young, the normal perils of childhood exacerbated by the dreadful conditions and acute privations of the missionary life. Gottlob, a man of guileless love and fervent dedication, was a pronounced worldly failure even there: falling into disgrace with distant authorities through various infractions, he had to abandon mission after mission, dragging his family behind him. They managed with no roads, no sanitation, and often no home at all, the family living in wagons or in other people's outbuildings. There are tales of floods and of other catastrophes, and occasional mention of Rebekah's suffering from what Gottlob identified as "inflammation of the brain." Gottlob was eventually stripped of his missionary status altogether, but at the time of Olive's birth, in 1855—the ninth child, with three already buried—the Schreiners had found temporary peace at a station in Wittebergen, on the edge of Basutoland. Even here Rebekah had a devastating range of household tasks to perform: the cooking, the sweeping and whitewashing of rooms, the making of clothes and of shoes, the rearing and basic education of the children. Olive Schreiner remembered Rebekah walking wearily out into the African twilight with a book under her arm when the children were finally put to bed. She envisioned her mother—"with her French and Italian, her flower painting and music"—as a grand piano closed up and locked, and tragically mistaken for a common dining table.

As a mother, Rebekah was cold and strict, having charged herself with the perfect preservation of London drawing-room gentilities even in the rudest mud hut—indeed, there more than anywhere, as insulation from the "gross sensual heathen" (Rebekah's words) among whom the family lived. It wasn't only black Africans' example that had to be repelled. Olive Schreiner's most searing early memory was of a beating—fifty strokes with a bundle of quince rods—that she received from her mother at the age of five for the offense of exclaiming "*Ach,* how nice it is outside," the meaningless exclamatory *Ach* being Dutch—that is, Afrikaans, and therefore vulgar. English was the sole language permitted in the household. (Olive's German father, it may be noted, was self-consciously ill at ease with English all his life.) "The bitter wild fierce agony in my heart was against God and man," she wrote of the beating many years later, and she clearly recast it as Waldo's faith-breaking ordeal in the fuel house, although even all-suffering Waldo is spared the lesson that unmerited pain may be delivered by the most righteous and beloved of hands.

Schreiner wrote—in letters—of only one other beating, meant to quell her self-will: she had stepped into the doorway to catch raindrops in her hand after being told to remain inside. She noted, however, that the two beatings "did me such immense harm that I think they have permanently influenced my life," and explained, "They made me hate everything in the heavens above and on the earth beneath." But the girl's wild anger was subverted by an intense and almost equally unbearable sympathy—a sympathy for whatever suffered—which was essential to the growth of the woman's broad ethic of compassion and her politics of emotional socialism.

That sympathy only increased with a tragedy that further challenged the credibility and the succor of conventional belief. "The most important event of my childhood," Schreiner wrote, was the birth of a baby sister, Ellie. "My love for her has shaped all my life." The love long outlived the child, who died at the age of eighteen months, when Olive was nine years old. It was this small death that, Olive recalled, "first made me realise the falsity of what I had been taught and made a freethinker of me."

By the time Olive Schreiner wrote about this loss, thirty years after the experience, she admitted to feeling none of the Heaven-storming enmity that marks the diaries and letters of other nineteenth-century women of religious conscience—women like Annie Besant, who had been a minister's pious wife—whose faith was snapped through witnessing the suffering or death of a child. By contrast, Olive Schreiner recalled the death of her sister as an experience more akin to a giving than to a taking, as the revelation of something larger and deeper than could be contained by the creed she had been taught. Bereavement carried her once and for all beyond the orthodox presumption of a select and exclusionary salvation. In a letter of 1892 she replied to questions about her religious faith, posed by an evangelical minister on the brink of resigning his post, with the lesson of her sister's death: "I cannot conceive of either birth or death, or anything but simple changes in the endless existence: how can I then believe or disbelieve in Immortality in the ordinary sense? . . . I used to love the birds and animals and inanimate nature better after she was dead; the whole of existence seemed to me more beautiful because it had brought forth and taken back to itself such a beautiful thing as she was to me."

Schreiner's childhood thirst for transcendence and her budding moral intuition were nourished by volumes of Emerson and of Mill—she scorned romances and hardly admitted to reading novels—and, above all, by one book that seemed to explain fully the inner workings of existence itself: *First Principles,* by the English savant Herbert Spencer. Although this book had provided the same revelatory function for legions of readers since its publication in 1862, and although Spencer's theories of social Darwinism ("the survival of the fittest" is his phrase) were even more lastingly influential, he is perhaps best remembered today as the dyspeptic prig who won George Eliot's heart and returned it untouched, fathoming neither its depths nor its value. His Casaubon-like life's achievement, the *Synthetic Philosophy,* in ten volumes (of which *First Principles* was, naturally, the first), accounted for all phenomena—from the solar system and nature to human thought and culture—through patterns of evolution based on the sharing and redistribution of matter and energy. The man who swore to George Eliot that he never frowned because he was never puzzled, had a formula—the same formula—for everything.

While hardly among the first to offer the ideal of Progress on earth as the thinking person's replacement for Heaven after death, Spencer was one of the few who thought the offer guaranteed: fulfillment was a matter of time, the world was on course. This was freethinking with a net, and it wasn't the only one Spencer offered, or even the largest one. For the ball of Herbert Spencer's universe was set rolling toward perfection by an all-purpose God-who-would-not-speak-His-name, who exists in Spencer's terminology as the Unknowable: "an Infinite and Eternal Energy whence all things proceed." The First Principle itself stated that all that was objectively knowable—the contents of all those coming volumes—rested upon "an Absolute that transcends not only human knowledge but human conception." Here was a tower of reason founded on a cushion of cloud: not only were religion and science not incompatible; they were hardly distinguishable.

It was *First Principles* that saved the teenage Olive Schreiner from furious atheism and despair and enabled her to write what her contemporaries called "an agnostic novel"—a designation attesting to the value of the new-minted word as a compromise between belligerent denial and the solace of belief. *First Principles* was given to Olive Schreiner by the first "freethinker" she ever met, a young man called Willie Bertram, who passed through her life for only a few hours on a tour of inspection for the Native Affairs Department. It became, accordingly, the mysterious book that the Stranger gave to Waldo "round which to hang your ideas"—a fact Schreiner volunteered to various friends; the book is named among the small collection that Waldo takes away from the farm. *First Principles* was for Schreiner what *The Way of a Pilgrim* was to her twentieth-century sister Franny Glass: a spiritual how-to for sneaking past reason and up to God. (*African Farm,* of course, became just another such book of the famished soul's last resort.) Willie Bertram committed suicide at the age of thirty-five, and even Herbert Spencer was reported to have said, toward the end of his days, "I have passed my life in beating the air." Olive Schreiner was told, after Spencer's death, that during the old man's final illness he had asked to have read to him the allegory of the hunter seeking Truth from *The Story of an African Farm.*

In 1881, when Olive Schreiner was twenty-six and had been working as a governess on a series of isolated homesteads, she sent off an application for nurse's training to the Royal Infirmary in Edinburgh.

> Religion: Freethinker—Father Lutheran
> Weight: There are no scales here
> Where Educated: At home

Through diligence and ingenious methods of forestalling sleep, she had already written two novels and several stories, in full or in part; a version of *African Farm* had been sent off to a publisher the previous year and returned with a list of suggested revisions. Longing for escape, she now envisaged a medical career. In later years, when her invalidism was well established, Schreiner claimed alternately to have been a sickly or a robust child, depending on the effect desired: lifelong suffering or dramatic collapse. She dated the onset of her asthma to a wagon journey of some four days' duration, made in cold rain and without food, when her father had lost yet another position and she—in early adolescence—was sent from her parents' home to live with an older sister. It was not the trip itself but the attempt to eat again, upon arrival, that brought on acute pains in her chest and stomach, pains that were to plague her all her life.

The voyage to England, en route to Edinburgh, was difficult for her.

Arriving in March 1881, she wrote in her journal, "I have not much hope. I shall *never* be well." In the nursing program, she lasted three days; an older brother, Fred, was called north from his home in England to rescue her. After recovering that summer under Fred's protection—throughout her years in England, her "Dadda," as she called him, provided financial support and, in times of crisis, a physical haven—she began a course of training at the Women's Hospital in one of the poorer sections of London. There she lasted five days before "inflammation of the lungs" forced her withdrawal. Another period of rest was sponsored by Fred, following which she began to attend lectures at the Women's Medical School. "The second lecture I went to I got my feet wet and sat in wet boots and got congestion of the lungs," she wrote home. "I am very strong and well now, but have made up my mind that scribbling will be my only work in life."

In the winter of 1882, Schreiner carried her revised manuscript of *African Farm* from publisher to publisher. It was accepted, finally, by no less a figure than George Meredith, an author himself well practiced in matters of the woman's predicament, his most recent heroine having been nearly married off in exchange for a cellar of port; two years after *The Egoist,* Meredith was working as a reader for the firm of Chapman & Hall. The publisher's only request of Schreiner was that the pregnant Lyndall be allowed a lawful wedding—a mere two or three lines—in order to placate the railway booksellers. The request was, of course, hotly refused. *The Story of an African Farm* was published, in two volumes, in early 1883, its spine embossed with a gold ostrich and its cover with a bit of desert landscape. The newly celebrated author was then living in a London boardinghouse, almost without acquaintances. She had spent much of the past two years—by her own account—in crying and in administering to herself large doses of potassium bromide, intended to suppress the action of her nervous system.

This constricted life opened out with a letter of admiration from a young medical student and aspiring literary critic named Havelock Ellis. *African Farm* had meant a great deal to him, he wrote, and he remarked on its affinities with the works of Thomas Hardy, whom Schreiner had not read. Ellis, four years younger than Schreiner, was a man of delicate health and mild but steadfast disposition. He had lost his religious faith in adolescence and sought palliatives similar to those found and now offered by Schreiner. (In the first bloom of his crisis, Ellis had written to Herbert Spencer but was discouraged from discipleship by a form-letter response.)

Ellis was soon introducing Schreiner into the progressive London circles of utopian societies and dream-ridden fellowships. She joined the Fellowship of the New Life—Ellis was a founder, George Bernard Shaw an early and mostly irate member—which was committed to the advent of socialist politics and sexual equality via "the cultivation of a perfect character for each and all." The overwhelming unlikelihood of achieving this required first step—a pure and unspotted soul demonstrated by every member—rendered any practical outcome of the fellowship's efforts virtually unthinkable. Mutual self-examination came to seem not its means but its goal, attained in an atmosphere of earnest Socratic titillation.

Havelock Ellis was himself interested very little in politics and very much in sex. Among the earliest "scientific" explorers of sexuality ("Perhaps some of the ideals you have realized in life have also been mine," Freud gracefully acknowledged), he meticulously studied, interviewed, catalogued, classified, and duly published, in many volumes. Ellis might be called the Herbert Spencer of sex: all-encompassing; immensely popular and useful in his time (his was a strong voice for tolerance); soon outdated and unread. There is another parallel: as Spencer elaborated theories of life in isolation from any contact with mere vulgar living, so Ellis became the paramount English authority on an act that he himself was for long years—for his full lifetime, according to his most recent biographer—incapable of committing. His enduring virginity perhaps lent him a certain objectiveness in his field of endeavor.

Olive Schreiner was the first woman Ellis loved. She taught him—as he said, gratefully—"to be *able* to long." He enabled her to fulfill her interest in medicine by encouraging her in her role of perpetual patient. But Schreiner's steady debilitation was as much the result of her needs and maneuvers as they were of his. In her letters to him she shifts from discussions of books and literary techniques to feeble cries of pain when she suspects that she has angered him; ever after, his sympathetic ear brings out the fullest play of her every wretched symptom. Ellis was tender, solicitous, devoted, and able to prescribe and deliver drugs: the potassium bromide she was already taking, and also quinine, chlorodyne, and, on occasion, opium pills and morphine.

As a pair, they struggled on in perpetual joint opposition to what they helplessly were, to what their times had made them. He revered her intellect; he suggested that thinking too hard brought an excess of blood to her head. She recounted every aspect of her sexual experiences to him, and he maintained dutiful records of such information as her longing for a past lover to tread on her and stamp her "fine into powder." Neither took cognizance of the difficulties such longings might cause a mind that was set on a hard and tenuous independence. She aimed at seducing him, and although they were never "what can be technically, or even ordinarily, called lovers"—in Ellis's words—they were intimates up to that point, and perhaps, as Ellis would have maintained, beyond it. Evidence of both the physical extent and the temperamental limits of their relationship is offered by one of Ellis's slyer comments in recalling a few days' idyll spent together outside London. He had brought along his microscope for her use, he wrote. "She wished to observe living spermatozoa, which there was no trouble in obtaining."

For Schreiner, the relationship was insufficient. She continued to keep Ellis abreast of her multiplying symptoms, but she drifted from his day-to-day activities, leaving the Fellowship of the New Life for a rebel offshoot, the Men and Women's Club. There, between lectures on the sex lives of Periclean Greeks, medieval Russians, and early Buddhists, the man with whom she soon fell hopelessly in love—an academic mathematician, Karl Pearson—and the man who fell just as hopelessly in love with her—Dr. Bryan Donkin, the Marx family physician—could joust in her presence on the question of the existence and the degree of the sexual instinct in women.

Olive Schreiner signed one of her letters to Karl Pearson "your man-friend"—letters to Ellis were often signed "your little sister" or even "your little child"—and far from complaining of her pains and inadequacies, she strove to produce for him an air of brisk competence and intellectual readiness. She had written little and published nothing since *African Farm.* To Pearson she now, in 1886, outlined a plan for a comprehensive study of woman's place in world society, to be based on "scientific" data and to range from cave life through Greece and Rome to the future. The history of women had been one of Pearson's strong private interests, and she may have hoped for a collaboration. Instead, he pronounced her the better man for the job and handed over the notes of several years' preliminary research. She suggested other collaborations. She became jealous of the women in his life while claiming not to be one herself. He rejected her or withdrew from her, and she broke down completely. Poor Dr. Donkin found her "in a state of complete temporary madness" and tried to intervene, encouraging Pearson to come to see her.

If Pearson came, he didn't stay long. Mortified, Schreiner wrote him a letter of staunch denial. "I never thought you loved me as a woman. You are drawn to me intellectually and I am of great interest to you." Donkin's interpretation of her breakdown had been mistaken: "You will forgive him. I do." She sent Pearson a final gift: her childhood Bible. Then she packed her bags and left the country.

Schreiner spent the better part of the next three years travelling in Italy and in Switzerland, continually on the move, constantly in pain, sending Ellis agonized letters with pleas for drugs: "What is given for hysteria? (Awful sensitiveness in lower part of body.) Isn't valerian a good thing?" For outsiders she could still summon an appearance of daunting, almost pugnacious strength. Even so dedicated and influential an agnostic as Leslie Stephen found her "a desperate freethinker," and also "hard and conceited." (Such "conceit" seems to have been a specifically female affliction, almost a clinical one. When Stephen's own daughter Virginia, aged fifteen, showed symptoms of hysteria—anorexia, depression, irrational fears—some two years after the death of her mother, the London specialist who was called in pointed to an excessive level of education and corresponding mental activity; more than a decade later he was treating the now adult Virginia Woolf for the same problems. "The danger of solitary work and want of social friction may be seen in conceit developing into insanity," he wrote.)

Schreiner labored in weary confusion at her second novel and—perhaps—at the big projected history of women, but all she managed to produce was a stream of symbolic "Dreams," brief allegories on her old themes. She claimed that they came to her whole, unaided by conscious thought; they could not be criticized because they were not "written." Published individually in London newspapers and then in 1890 as a collection, the "Dreams" met with success, championed by the likes of Shaw and Oscar Wilde. They seem today at once affecting and almost comically overwrought. A flat, oracular tone replaces the youthful breathlessness of *African Farm.* God has emerged from hiding with a quantity of epigrammatic reassurances: "In the least Heaven sex reigns supreme; in the higher it is not noticed; but in the highest it does not exist." The sole poignant note in this little story of the three Heavens, in which a single creature rises to the highest sexlessness, comes in Schreiner's slipping from the pronominal "it," with which the narrator remarks the risen creature, into God's own careless use of "he." The battle within was always the hardest one; she was as exhausted as she claimed to be.

▪ ▪ ▪

In late 1889 Schreiner returned to southern Africa—not a country, and no longer even resembling a colony so much as a British-controlled corporation overseeing diamond and gold production. In London, she had come to a sense of political futility—all that talk changing nothing. Now she was a great celebrity in a small and volatile society, and her opinions carried force. These were slow, however, in developing. For a long while, she rebelled against the trespasses and the materialism of the British by romanticizing the scrubbed and sturdy farming families of the Boer republics, who had come to Africa to live and work the land, not to rob it and flee. She saw no other claims, and was long considered a powerful friend—later, even an apologist—for the cause of the white Afrikaners.

Despite this partisanship, Schreiner was captivated by the British empire builder Cecil Rhodes, even comparing that ruthless archexploiter with her own dreamy Waldo. (Rhodes adored *African Farm.*) He was to her "the only big man we have here." Schreiner stood in furious public opposition to Rhodes over his parliamentary support for the so-called Strop Bill, which authorized flogging and imprisonment for workers in the mines. But a bond of personal affection is still apparent in a skit that she wrote: Rhodes is sent to Hell but turned away because he is too large to fit through the gates; a special place must be found for him upstairs. Schreiner's first stirrings of concern for the native African population—seldom noticed by her before, and then only in the gross stereotypes of her childhood—seem to have begun at this time, in a surge of empathy brought on by her own memories of being whipped. Of her rediscovered homeland she writes to a friend, "There are money-making whites and downtrodden blacks and nothing in between. And things will have to be so much worse here before they can be better."

In December 1892 Schreiner met another would-be Waldo. Samuel Cronwright was a young farmer—eight years her junior—who had given up thoughts of a church career when beset by religious doubts. By the time he read *African Farm,* two years before meeting its author, Cronwright felt he had already lived through the book's spiritual struggles. He emerged from Schreiner's work "more than ever a Freethinker," newly impressed by "the imperativeness of mental truth." He vowed that if he ever met Lyndall he would marry her.

The real Lyndall, when he did meet her, struck him as very strong, very bold, fearlessly opinionated, "just life and force and brilliancy personified." She liked him, too. Her first letter after their meeting assured him that "a short description of the ostrich and its habits, as you would write it, would be of real value." To this end, she offered the literary services of her agent in London. She sent Cronwright some of her allegories, making certain to correct any false impression that might arise: "I do not think the celibate life the ideal." Cron, as she called him, was soon sent off to meet Rebekah, now a widow and retired to a Roman Catholic convent. The telegram that mother sent daughter leaped back more than fifty stark years to recall a bride with forbidden flowers on her hat: "Called. Feast reason. Mental champagne."

Cronwright was no intellectual gentleman farmer. He was a man of almost vainglorious physical strength, who sweated proudly over his eleven thousand acres, who bragged of sitting bucking horses and hefting hundred-pound bags of grain—a single-bullet leopard-and-antelope man. The letter Rebekah sent after her telegram fairly cooed over "his physical manliness, his muscular prowess"—perhaps one should recall that it was Cronwright who published the letters—and Olive requested a photograph of him in rolled-up shirtsleeves to take with her on a trip back to England.

Her reasons for the trip are not quite clear. She sailed a day before her younger brother Will was sworn in as attorney general of the Rhodes Cabinet, but her plans were made well before that, and her mind was not on politics. A clue may lie in a story she wrote before she left and which she immediately sent to Cronwright: a retelling of her disastrous parting from Pearson, in which a nameless and cynical hero awakens to the love of a nameless, wise, and infinitely self-controlled heroine just as she vanishes forever, from the room and from his life; she has gone off to India, alone, seeking knowledge. Schreiner penned a farewell to Cronwright—"my brother, my boy"—and informed him of a plan to settle down in Europe to finish one of her "big books." It is likely that he had proposed to her, and that her resistance, as with her breakdown over Pearson, was both genuine—she desperately required the financial independence another book would provide—and contrived. In her journal she muses over whether or not he will follow. In England, she stayed among friends, and her writing was confined to long letters back to Cronwright. By August the marriage was set.

Set—and nearly destroyed by a confession that Schreiner apparently made immediately upon her return to the Cape that fall. Cronwright's coyly edited letters leave us such hints and elisions as "Cron, it was all my fault about X. . . . Your Olive was so weak and she's so ashamed." Without direct accusation Cronwright makes it easy to suppose a sexual transgression, although none of her subsequent biographers have in fact so supposed; these letters have been either ignored or consigned to ineffable mystery. Circumstances would have been especially opportune during her stay at a free-spirited ashram that an old friend, Edward Carpenter, had set up in Derbyshire, a kind of protohippie commune where city bohemians and Oxford dons mixed with the brawnier class of local quarrymen and colliers—the place where E. M. Forster was to be literally touched into writing the sexually explicit *Maurice* by a hand placed on his backside. (The hand belonged to George Merrill, Carpenter's longtime lover. "I believe he touched most people's," Forster noted.)

Once back in Africa, Schreiner was too sick to rise from her bed. Cronwright saw her once and then seems to have turned his back. She sent word that she could not walk, could not speak. She wrote letters of supplication and quivering self-abasement, begging that her "wickedness and weakness" should not shake his ideal. In *African Farm* Lyndall sneers at woman's ever-present easy way out: "A little weeping, a little wheedling, a little self-degradation, a little careful use of our advantages, and then some man will say—'Come, be my wife.' " Lyndall was seventeen; Olive Schreiner was now thirty-eight. She and Cronwright were married, by civil ceremony, on February 24, 1894.

The marriage wasn't a giving up. It was, on the part of both, a difficult effort to meet a stern ideal. He took her name, becoming Samuel Cronwright-Schreiner, and she kept her own: she felt herself already too well known for anything else. They signed an antenuptial contract that kept their properties and incomes separate. In marriage, a woman had to be "absolutely and entirely monetarily independent of the man," she wrote to a friend. "That is the great thing; let love bind you, not a common account in the bank." The fact is that her writing brought her very little money. With the steady aid of her brother Fred now concluding in a dowry gift, she would continually fall back on Cronwright's benefactions and on carefully accounted and unrepayable "loans."

The newlyweds settled down in happiness on his farm. Within weeks, she found that she could not breathe; there were terrible asthma attacks every night. She said the siting was bad. Cronwright felt he had married a genius ("In Olive I had a sacred trust"), and he agreed to give up his farm and move with her to a place where she could live and work in peace. He did not yet know that there was no such place. She knew it—as early as 1885 she had written to Ellis, "It isn't my chest, it isn't my legs, it's I myself, my life"—but her knowledge was intermittent, pushed aside, unendurable. She asked for two years on a homestead that she bought for them—he would be without occupation there—in order to complete the "big" novels she was working on; she aimed at making them both financially independent. In the four years they went on to live there, Schreiner did not write any "big books" but, rather, political pamphlets and articles and speeches, often in collaboration with Cronwright. She had a baby—her only one, a girl—who lived only sixteen hours. The child was buried in their garden. When they moved, the coffin went with them.

In 1896 Schreiner completed a long story, a political allegory based on Cecil Rhodes's most recent and brutal retaliations against uprisings of the Matabele and the Mashona, long residents of the state that was to become Rhodesia. The hero of her title, "Trooper Peter Halket of Mashonaland," is a softish English soldier in Rhodes's company, not much more than a boy, whose routine racial and cultural presumptions are tinged with revulsion at the beatings and hangings and shootings carried out and often enjoyed by his companions. Alone one night at a desert campfire, he is visited by a barefoot man of mercy, a Jew from Palestine (" 'Have you ever been without grub?' said Peter cheerfully. . . . 'Forty days and nights,' said the stranger"), who wins his allegiance. The next morning, Peter cuts loose a native prisoner and is summarily shot by his commander. *Trooper Peter* was issued on its own, in 1897, as a novella. Its frontispiece was a photograph of a large spreading tree in the African landscape, with eight white frontiersmen calmly disposed beneath it and three black men hanging from its boughs.

The Cronwright-Schreiners travelled together to London in January 1897 to see to the new work's publication. It was during this extended visit that both Canon Wilberforce and a Wesleyan minister, Hugh Price Hughes—encouraged, doubtless, by the materialization of the figure of Jesus in her narrative—paid separate calls on Schreiner, appealing for her spiritual force. *Trooper Peter* was, as could have been predicted, a polemical bombshell. *The Spectator*'s review suggested that it might have been another *Uncle Tom's Cabin* but for Schreiner's no longer acceptable inflated rhetoric. In New York, the *Tribune* blasted, "Let the English explain away, if they can, the murder and rapine which Olive Schreiner lays at their doors." As for Schreiner, she spoke of *Trooper Peter* sometimes as her most important work, at other times as a complete failure: it didn't save a single life. The unnerving photograph was removed from later editions. When the novella was published in South Africa in 1974, the frontispiece was finally restored, but with the plate tipped in for easy removal in case of censorship.

For Samuel Cronwright, *Trooper Peter,* despite its impact, was not what she had promised—not one of the "big books" she had set out to write. Indeed, "she did not seem to remember that I had taken the grave step of giving up my livelihood at her special request." If Wilberforce had expected Olive Schreiner to be another St. Teresa, Cronwright had believed she could become another George Eliot. She aroused the expectations; she lived in their daily smashup.

Cronwright took back what he could of his life. There was no money for a new farm, but he trained for the law in Johannesburg and eventually became a businessman and a member of the Cape Parliament. He and Schreiner shared a political life, campaigning against the outbreak and the continuation of the Boer War. In 1900 Cronwright went alone on a six-month speaking tour of England to rally antiwar support. Other years, they were apart for similar lengths of time, as Schreiner would go off for months—usually to the coast during the hot African summers—in search of air and calm. During these periods, she wrote to Cronwright constantly, letters of vague politics and insistent devotion. By 1906 there were marked changes. A suffragist friend had given her W. E. B. Du Bois's *The Souls of Black Folk,* and she wrote to Cronwright that she no longer cared for the Boer cause: "They are more than able to take care of themselves. But the Natives are always with me." To her old friend Ellis she wrote brightly of her continuing faith in the greatness of human nature, "though my personal life has become crushed and indifferent to me."

The rights of the native Africans were now annexed to her political vigilance as a cause second only to the rights of women. In a country with a color bar, these were not distinct issues: a founder, in 1907, of the Women's Enfranchisement League, Schreiner resigned when its objective was formulated as the winning of the vote on the same terms as it was held by men in southern Africa. She would not cut the moral knot, would not abandon women of color, no matter how much longer—how impractical—the resulting wait for all.

In these years, Schreiner completed all she ever would of the "sex book" proposed to Pearson decades earlier. The tract, called *Woman and Labour,* published in London in 1911, reestablished her reputation as a social prophet. The purpose and vigor of woman's long history of necessary labor, she argued, had been replaced first by the use of slavery, as in ancient Greece, and now by modern industry, and each had transformed woman herself into "the 'fine lady,' the human female parasite—the most deadly microbe which can make its appearance on the surface of any social organism." In the modern age, Schreiner wrote, cultural decline was sure to follow from the passive condition of idle women, whose marriages were less partnerships than prostitutions. (Schreiner had evidently read Engels on the origins of the family.) As mothers—and Schreiner revered motherhood—women shaped the entire race with each new generation, for better or for worse. The parasitism of the childbearing woman, a kind of mental gout, could lead only to "the enervation and degeneration of a class or race." With women's traditional work at the spinning wheel and grindstone and hoe securely eliminated, the solution to this dilemma was the book's great battle cry: "We take all labour for our province!"

Schreiner's book was received by many as "the Bible of the Women's Movement." Forty-five years after John Stuart Mill first introduced a parliamentary petition for woman suffrage in England, seven years before the vote was won, the plain message of *Woman and Labour* still hit like a news dispatch: there was no work or profession for which a woman, simply by her nature as a woman, was unfit. "There is no fruit in the garden of knowledge it is not our determination to eat," Schreiner maintained, and her book influenced many a decision to go to university or to otherwise seek out a life apart from that prepared for the English young lady. D. H. Lawrence set his Ursula Brangwen under a hawthorn tree during a school-dinner break with a copy of *Woman and Labour* and all her longings for "some other life besides . . . housework and hanging about," her bold desire even "to earn something."

The narrowness of the book's range—its concentration on the relatively limited problem of contemporary middle- and upper-class "parasitism"—was due to its nature as a fragment of a larger work. In her introduction Schreiner sketched the whole as it would have stood: chapters on lower biological forms and on women in the "savage and in the semi-savage state"; physiological studies to explain the pain of childbirth and to suggest that woman—not man—first rose onto two legs (under the necessity of carrying her young while seeking food or escaping from enemies); a concluding investigation of the new women's movement and its consequences for society. All this near lifetime of work, she explained—twelve chapters carefully researched and analyzed, and typewritten and bound—had been burned in a fire set to papers taken from her desk when her Johannesburg house was looted by British soldiers during the Boer War. She had not been living in the house at the time, but in a small town up north, where she had been trapped at the war's start. It was a year and a half before she returned to find her manuscript book and its pages "so browned and scorched with the flames that they broke as you touched them; and there was nothing left but to destroy it." She lacked the strength to go back and rewrite; the current work was offered only as a small and regretful remembrance.

This whole story Cronwright later called a lie. There had been no research, he claimed, no analysis, no book, no consuming fire. No papers of consequence had been left—had ever been left—at such a distance from where she was living. The house had indeed been looted, but at the time she had voiced regret at the loss of no more than a box of personal treasures: a lock of her father's hair and of her baby sister Ellie's, some old journals, and a sheaf of "Dreams." Moreover, "such a book would have meant hard, exact, systematic reading and study, and a collection and tabulation of exact scientific facts, a kind of labour she was incapable of." She constantly confused the terms "thought" and "work," Cronwright said. "She may have dwelt upon the thought of this imaginary sex book until, for her, it assumed objective form." She was, in short, a pathetic monster of self-deception.

In Cronwright's *The Life of Olive Schreiner* and in his edition of her letters, both published in 1924, four years after her death, the oppressed husband got his own back. These are jarringly nasty but seemingly rather honest books, their official fabric of adulation riddled by snorts of derisive laughter and indignation. Schreiner no more than mentions in a letter that she has been playing Chopin on the veranda but a bristling footnote informs us, "She could not play, only strum simple tunes like a child." He catches her out in not having finished reading a book she claimed to admire; he reproduces the self-devised study schedule of her young governess days—Latin, French, mathematics—and declares that it is "worthless, except as her categorical setting down of what was proposed to be done, and then not doing it, a way she had almost to the last." The letters themselves—a few hundred chosen from over six thousand—were "a justification for postponing her more arduous literary work."

Schreiner sailed for England and Italy in 1913, seeking help for a heart condition that now barely allowed her the strength to move. She didn't have the money for the trip and tried to barter a manuscript to manage it, but finally allowed Cronwright to pay her way. (When she allowed him to assist her like this, he relates, "she said playfully she was becoming a 'parasite.' ") They were separated for the next six years. She was in Germany, taking the waters, at the outbreak of the Great War and succeeded in making it back to London. There she remained for the duration, a staunch pacifist who could hardly get a room in a boardinghouse because of her German name. She broke with most of the suffragist movement for its fervent support of the war, and argued with Gandhi about his recruiting an Indian ambulance troop for Britain. Poor as she was, she refused to allow a film to be made of *African Farm:* "The tragedy of Lyndall and Waldo can't be put into a picture because it is purely intellectual and spiritual."

She and Cronwright wrote to each other constantly. In June 1920—she was still in London—he said in reply to one of her letters, "When you write like that, my heart is water. . . . I love, revere and reverence you, and have loved, revered and reverenced you, as I have never done anyone else." He arrived in London the following month to be with her; he later said that he barely recognized the old lady who greeted him. Olive Schreiner sailed back to South Africa in August. She spent her last days in a boardinghouse in a Cape Town suburb, soliciting money for the defense of an African National Congress strike organizer and hoping to become strong enough to go out to the karroo one more time, or to find someone to take her there. She died in December. She was sixty-five years old; an autopsy—she had requested one, as an aid to future sufferers—revealed blocked coronary arteries, accounting for her continuous anginal pain. Her lungs showed one long-healed scar, very small, and were enlarged with emphysema. There was no attending minister at the funeral held a few days later, and the ceremony was carried out in complete silence, to prevent any of the religious sentiments so common at gravesides from being spoken.

By the time Virginia Woolf reviewed Cronwright's edition of Schreiner's letters in 1925, Schreiner was a "rather distant and unfamiliar figure," despite having so nearly become, as Woolf noted, "the equal of our greatest novelists." Withal, she had failed: she remained "one half of a great writer." In the year of *Mrs. Dalloway,* with its teasing nod to "women's rights (that antediluvian topic)," Woolf complained of Schreiner's unrelenting obsession with "questions affecting women." Four years later, Woolf wrote *A Room of One's Own,* and feminist critics have pointed out an important link with Schreiner's second novel, *From Man to Man,* the unfinished labor of more than a quarter century, finally published just two years before Woolf's famous essay. Schreiner had begun her book on the day she finished reading Darwin's *The Descent of Man,* in 1873, and her thwarted tale of sisterhood centers on a female counterproposal to the theory of natural selection: survival as the result of self-sacrificing mother love. Woolf's all-important room belonged first to Schreiner's theorizing heroine Rebekah, who brought it into being "by cutting off the end of the children's bedroom with a partition." Used as a study and filled with Rebekah's manuscripts and books, her scientific collections and her microscope, it was the only room in the house kept locked.

Virginia Woolf assumed that Schreiner's work would disappear, and for many years, outside South Africa, it did. In her homeland, there have always been champions, perhaps because there have always been those who were given *African Farm* to read when young. Both Doris Lessing and Nadine Gordimer have written about Schreiner with great feeling; Lessing has acclaimed *African Farm* as one of the small number of books existing "on a frontier of the human mind," in company with *Moby-Dick, Jude the Obscure,* and *Wuthering Heights.* Both authors may trace their descent from Schreiner's example, not only in the literary depiction of a real South Africa but in a fiction of large ideas and political reverberations. Isak Dinesen read *African Farm* at the age of fifteen, before she had any idea of a life in Africa, and although the viridian-lush hills and polished leaves of Dinesen's East Africa bear little resemblance to Schreiner's arid biblical testing grounds, Dinesen wrote of remembering and recognizing, years later, the "high, incessantly varying" African air and wide sky as Schreiner had promised them, and of recognizing, too, Schreiner's perception of the land's essential tragedy.

Schreiner has a new audience today—responsible for a considerable revival in her fortunes—among English and American literary feminists: Lyndall's children. In *The Feminine Mystique* in 1963, Betty Friedan paid tribute to Schreiner's early warning about women's loss of meaningful work. Tillie Olsen's melancholy *Silences* of 1965, a book about books unwritten, offered sympathetic attention and entered Schreiner's name on a list of women writers ominously titled "Not Suicides," with the implication of Not Quite. Then, in 1977, Elaine Showalter's groundbreaking *A Literature of Their Own* seemed to fire a starting pistol by declaring Lyndall "the first wholly serious feminist heroine in the English novel." That year, Cassandra Editions reissued all of Schreiner's major fiction, including *African Farm* and *From Man to Man* and a collection called *Dream Life and Real Life;* a year later, the Virago Press gave *Woman and Labour* its first reprinting since 1911. From there, it was a short step to *An Olive Schreiner Reader,* published by Pandora in 1987 with a dedication to Winnie Mandela, and to a rapidly accumulating bibliography. Olive Schreiner was being turned from an author into a heroine.

Even so, it was as difficult for some modern feminists to forgive Schreiner for defrauded expectations as it had been for Cronwright. Despite the historical importance that Showalter bestowed, she declared Schreiner's two novels "depressing and claustrophobic." Schreiner was "sadly underambitious"; her heroines "give up too easily and too soon." The difficulty with taking Lyndall as a feminist rallying point had been addressed as far back as the fellowship drawing rooms, where Schreiner's friend Elizabeth Cobb remarked skeptically that people wanted to be "helped to move, not to lie down and die." But Schreiner could as easily have allowed Lyndall to sail off into a bright future as have married her off at the publisher's request; it would have falsified all she knew. There was no future imaginable, yet, for Lyndall or for Waldo.

And as for that God-lusting boy, Schreiner's other half, so nearly forgotten in the current reconsideration, it is disconcerting to find the shrewd and fluent Showalter writing that Waldo, at the last, "becomes an intellectual, and finds peace in transcendentalism," ignoring the fact that what he actually does is die. Schreiner tried to be subtle, but her meaning is clear, and references to Waldo's death recur in the letters. This curious omission allows Showalter to impose a theme on Schreiner's book: woman's suffering leads to man's redemption. Schreiner's contemporaries knew better; in her little book "as sad as life," no one is redeemed.

Schreiner's life and its implications have been written and readjusted many times. Cronwright's biography and his edition of the letters enraged her admirers from the start. Vera Buchanan-Gould's *Not Without Honor: The Life and Writing of Olive Schreiner,* the first full-length portrait to follow his, in 1949, shot back with statements such as "Like most men he was inclined to be impatient and unsympathetic with continual suffering." In 1980 several subsequent partial and amateur articles and books—Schreiner as martyr, Schreiner as latent homosexual, and so on—were swept aside by the publication of Ruth First and Ann Scott's measured and scholarly *Olive Schreiner.* Ruth First was a South African antiapartheid activist, assassinated in 1982, and Ann Scott is a British critic. Their collaboration benefits most from a finely detailed presentation of the social and political background of Schreiner's English experience and, particularly, of the struggling colonial Africa of her youth and middle age. The book is packed with information, most of it judiciously set forth. But the authors present Schreiner herself largely as a social and political figure, distorting the relative importance of her accomplishments. They demonstrate far less interest in the novel that is, after all, the reason Schreiner's thoughts and positions were solicited throughout her life. Glossing over elements of the story of *African Farm* as Showalter does, they misrepresent its intentions just about as thoroughly. Waldo's beating in the fuel house is missing from their retelling, and they seem rather flustered by Showalter's conclusion about his fate. "He falls asleep in the sun. Or does he die?" they ask. Moreover, much of Schreiner's book is distorted through political strong-arming to accord with the authors' avowed intention of placing her in a "Marxist-feminist framework." The rascally Blenkins's tall tale of a fortune lost at sea, which he uses as an excuse for his tattered appearance on arrival at the farm—an excuse that Lyndall sees through immediately—is accepted by First and Scott at face value, so that his bumptious schemes can be associated with "the entrepreneurial ideology of an emergent capitalism."

Olive Schreiner's feelings about native Africans—not her official late-in-life position but her frequently and publicly expressed feelings of condescension and aversion—offer a problem to anyone seeking to award her a medal for human or political enlightenment. Her most highly lauded political pamphlet, written in 1899 in advocacy of a democratic and integrated national union, offered as reasoning: "The dark man is the child the gods have given us in South Africa for our curse or our blessing." It is not surprising that Schreiner has been denounced in some quarters as a "racist feminist." First and Scott, sensibly aware of the climate in which she matured and the traps of moral hindsight, state simply that Schreiner was "firmly rooted in the racist stereotypes of contemporary ethnology which she was clearly unable to transcend." In their discussion of *African Farm,* however, the absence of native Africans as anything more than specks on the landscape or briefly glimpsed and unattractive intruders is transformed into "the point about the colonial conditions: Africans were kept so far outside white society that that in itself was a statement about it." Likewise, the brutishness of Bonaparte Blenkins and Tant' Sannie is but the effect of a colonial society that "internalized the violence" used against the indigenous population. Finally, the novel itself offers "a statement about the violence of colonialism." Yet just a few pages after this, First and Scott quote Schreiner as arguing that Lyndall's death was "meant to show the struggle of helpless human nature against the great forces of the universe."

The arbitrary cruelty and misery that Schreiner saw and recorded have their roots deeper than any strict political reading, Marxist or other, will allow.

Cronwright is as much of a villain in the new biographies as in the old, and the terms haven't changed much. First and Scott find him "insensitive to her struggles for expression either as a woman or as a writer." They believe him to be wrong or lying about the nonexistence of Schreiner's big "sex book," yet their accusation of "literalism over the manuscript" makes them seem as confused about the difference between thought and writing as Cronwright claimed Schreiner to have been, particularly when they add that he was "unsympathetic to her discontinuous, almost passive way of working." This way of "working," if that is what it was, cost Schreiner dearly, and her biographers' words condone the methods most derided by the woman who finally accomplished Schreiner's literary goal. "Woman is ready enough to play at working," Simone de Beauvoir warned in 1949, in *The Second Sex*, "but she does not work; believing in the magic virtues of passivity, she confuses incantations and acts, symbolic gestures and effective behavior."

Olive Schreiner saw herself, toward the end of her life, as equally important—as an example—in failure and in success. She knew that she suffered the ills and dissatisfactions of a "transitory condition," her own and her society's, and she consoled herself with Browning's "What I aspired to be and was not, comforts me." Her introduction to *Woman and Labour*, which she addressed to future generations, concludes:

> You will look back at us with astonishment! You will wonder at passionate struggles that accomplished so little; at the, to you, obvious paths to attain our ends which we did not take; at the intolerable evils before which it will seem to you we sat down passive; at the great truths staring us in the face, which we failed to see; at the truths we grasped at, but could never quite get our fingers round. You will marvel at the labor that ended in so little;—but, what you will never know is how it was thinking of you and for you, that we struggled as we did and accomplished the little which we have done; that it was in the thought of your larger realization and fuller life, that we found consolation for the futilities of our own.

Schreiner's resuscitation at the hands of the women's studies departments may not qualify as a second life of art, but it is a kind of sweet revenge. She who complained of not having had sixpence spent on her education found herself in tears at her nephew's university graduation, with its "great crowd of girls in their caps and gowns sitting among the boys." Her last testament included the request that if enough money should remain in her estate after a sum was given over to Cronwright, it should be used to found a scholarship for women at the South African College, Cape Town. To be called "The Cron Scholarship," the bequest would be available to women of all races and religions, with preference given to the poor. Cronwright concludes his account of Schreiner's life with the information that her estate was in fact financially negligible, and that he therefore took it upon himself to establish this educational trust. Spending only his own hard-earned money, as he reminds us, he fulfilled her proposal to the letter, departing from it only in the small matter of a title. Schreiner's fondest last wish and contribution to womankind was commemorated as "The Olive Schreiner Scholarship: Founded by her Husband."

The Mother of Confusion

Gertrude Stein

"Pablo & Matisse have a maleness that belongs to genius," Gertrude Stein scrawled in a notebook, sometime in her early Paris years, about the two painters she planned to join in re-creating Western art. She was an aspiring writer just past thirty when she met the paired geniuses, in 1905, and their spectacular audacity helped her to determine her own ambitions: to overturn the nineteenth century's constraining rules and prejudices, and to find new words for people's secret inner lives. Stein—a Jewish woman who'd studied psychology at Harvard—had fled America to join her brother on the Left Bank in 1903, to write an anguished novel while he painted mediocre nudes. It was at that moment, when modern art was being born, that the Steins began to buy the best and most shocking paintings; before long, the best and most unshockable people were stopping by to view their riotous, thick-hung walls—and were staying for dinner. Although Leo had begun the collection, Gertrude was the one Picasso wanted to paint and the one he called Pard, using some cowboy slang he'd picked up in American comic strips. Casting off her stays and starting a new novel, she began to think it possible that she, too, might be a genius—that she might do in words what Picasso and Matisse had done in paint—and if maleness was a necessary part of it, well: "*moi aussi* perhaps," as she added in her notebook. That was not a problem but an opportunity to demonstrate a truth of her own secret inner life.

The first decade of the new century was barely over before she had achieved a version of her goal: the first indubitably modern literary style. Before James Joyce—as she volubly insisted all her life—before Dada or Surrealism, before Bloomsbury or the *roman fleuve,* Gertrude Stein was writing books and stories that were formally fractured, emotionally inscrutable, and, above all, dauntingly unreadable. This achievement has given the Library of America a particularly difficult task in assembling two hefty volumes of Stein's selected works. How best to represent her legacy? It is often forgotten that Stein commanded a broad literary range, from the psychological realism of her earliest fiction to the journalistic accounts of life in occupied France during the Second World War. The work that made her famous and still earns her canonical status is, of course, the barrage of janglingly repetitive lyric obfuscation that has come to be known as Steinese. "A rose is a rose is a rose is a rose," she wrote, with the philosophic kick of a nursery Wittgenstein. "The sister was not a mister," she warned, out on a sexual edge, and then provided her own commentary: "Was this a surprise. It was." Although she did not write "Yes! We Have No Bananas," it isn't surprising that the song has been accused of betraying her influence.

It was by perpetrating such suspiciously significant nonsense, somewhere between the studies of Freud and the logic of the Red Queen—"Before the Flowers of Friendship Faded Friendship Faded" one cautionary title runs—that Stein entered our language as the bard of a culture of confusion, the vastly imperturbable mother of an age that had given up on answers. Yet no one took more vivid pleasure in the questions than she did, or set them out in a more brilliant company, beginning with the famous salon where she gathered Picasso and Matisse and Braque (who was so strong and so amiable that he would help the janitor hang the bigger pictures) and Derain and Juan Gris and Apollinaire. And after the Great War had blown this brilliant world apart, her rooms filled with charter members of the next cultural resurgence, and then the next, as there entered Hemingway, Fitzgerald, Cocteau, Tchelitchev, Christian Bérard, Cecil Beaton, Thornton Wilder, Virgil Thomson, and Richard Wright. For decades, there seemed no end to her gifts of renewal. She was host, sponsor, critic, instigator, frequently foe, and sometimes friend again of some of the century's finest provocateurs, and it was often hard to tell whether her life was a party or a revolution. But her intent was serious ("desperately" serious, as Alice B. Toklas put it), and she knew all along that the stakes were as high as the opposition was fierce, albeit nervous. "They needn't be so afraid of their damn culture," she erupted early on, and, for once, hung back in estimating her powers: "It would take more than a man like me to hurt it."

How did a girl born in Allegheny, Pennsylvania, in 1874 turn into such a disruptive fellow? Her grandparents had sailed from Germany to Baltimore in time to take opposing sides in the Civil War—a difference of opinion that she thought well established the Stein familial spirit. She was the youngest of five, although she seems to have felt herself to be the youngest of seven, so often did she dwell on two siblings who had died before she was born—an infant boy and a stillborn girl—and without whose deaths neither she nor her brother Leo would have been brought into the world. Both Leo and Gertrude had perceived some regret on their parent's parts at the quality of the substitution. This was one of the things that drew the youngest pair so close together, along with the knowledge that they were smarter than everyone else they knew.

The family moved from Allegheny to Vienna when Gertrude was a baby, and then to Paris (where she added French to her English and German) and, finally, to the wilds of Oakland, California, in 1880, when Daniel Stein forbade his children to speak any language but "pure" American English. A cold and domineering man of minimal formal education, he went on to make a small fortune by investing in San Francisco cable cars, and provided well for the children whose lives he overloaded with tutors and lessons and ambitions. Leo recalled his childhood as a torturous regime lived out under his father's disapproval, while his softhearted mother was too weak to make any difference; Gertrude would only allow that she considered it a foolish idea to have had an unhappy anything, let alone something as important as a childhood.

But there can have been little room for bluster back in 1886, when their mother began to show symptoms of abdominal cancer. Years of wasting pain and withdrawal turned Milly Stein into something of a household wraith before she died in 1888, when her younger daughter was fourteen. In *The Making of Americans,* the massive novel Gertrude Stein wrote about her family—the first stunningly original disaster of modernism, finished in 1911, written as though from deep within a halting, troubled mind—she suggests that Milly's illness had rendered her so nearly invisible that neither her husband nor her children noticed when she finally disappeared.

> Sometimes then they would be good to her, mostly they forgot about her, slowly she died away among them and then there was no more of living for her, she died away from all of them. She had never been really important to any of them. . . . Mostly for them she had no existence in her and then she died away and the gentle scared little woman was all that they ever after remembered of her.

Stein's notebooks, however, tell a different story: "All stopped after death of mother."

In fact, the household fell into chaos, while eighteen-year-old Bertha struggled to make up for a maternal presence that no one else would even acknowledge was missing. But if Gertrude had thought her mother negligible, she found her older sister sorely disgusting: "pure female," she wrote in her notebook for *The Making of Americans,* "sloppy oozy female . . . good, superior, maternal." Alone and untended, Gertrude left high school—her most thorough early biographer, Brenda Wineapple, states that she simply vanished from all records—and plunged into what she later called her "dark and dreadful" days. She took to reading with a kind of violence, buying piles of books (often with advice from Leo) and gulping them in great haphazard quantities. And she developed an almost equally violent need for food—"books and food, food and books, both excellent things," she wrote. Her discovery of food as a way back to early childhood, to "the full satisfied sense of being stuffed up with eating," marked her first divergence from the ways of her beloved Leo, who was just discovering how much he preferred to starve.

The case against Daniel Stein was set out many times by his two youngest children. They hated him, and yet they respected his power; they feared him, yet they wanted to be like him, if only for self-protection. More ambiguously, *The Making of Americans* contains some bizarrely dreamlike scenes that, among Stein scholars, have long raised the question of incest. (In the clearest example, a grown daughter accuses her father of having introduced her to an unnamed vice, and her words cause him to fall down paralyzed.)

In 1995 the biographer Linda Wagner-Martin, having studied Gertrude's notes for the novel, concluded that it was Bertha who had been sexually approached by their father: Gertrude wrote of his "coming in to her one night, to come and keep him warm." Complicating matters further, Gertrude elaborated by referring to a similar incident she had undergone with an uncle, her father's brother Solomon. "Scene like the kind I had with Sol," she wrote, and added, in a strangulated shorthand, "like me what he tried to do." Whatever actually took place, it is striking that just at those moments when she approaches what she cannot say, even in her own notes, this cautious writer begins to sound like quintessential Gertrude Stein. "Fathers loving children young girls," she goes on, and it appears that at least one branch of modern literary style may derive from a fearful evasion of meaning, and from the necessary invention—and wasn't that necessity "pure female"?—of a secret code.

"Then our life without a father began, a very pleasant one" was Stein's typically cloudless way of addressing the death of Daniel Stein—by apoplexy, three years after the death of his wife. Certainly his death precipitated a release of productive energy in his younger daughter. In 1893 she followed Leo to Harvard—this was a year before the Harvard Annex for Women became Radcliffe College—where, studying English and psychology, she turned out a quantity of notebooks in which her private turmoils were forced into traditional narrative forms. Her idols were George Eliot and Henry James, but her first work to be published was "Normal Motor Automatism," a psychological report she produced under the aegis of William James, who was doubtless responsible for rerouting her toward psychology as a science rather than, as in his brother's work, psychology as an aspect of literature.

In 1898 she went on to medical school at Johns Hopkins, planning to study the mysterious female affliction known as hysteria. Freud's book on the subject had been published three years before, but the frailties of women had preoccupied Stein since her childhood. Now she found refuge in the theory, not uncommon at the time, that a few rare women are born exceptions to their sex. To be extremely accomplished or intelligent or determined—to aspire to genius—was to be, by definition, less female. Although a friend persuaded her to present a public paper on the benefits of women's education, the last thing she was interested in was the cause of women's rights. As far as she could see, the only thing most women excelled at was having babies.

She did well in her first years of medical school, when the work centered on the classroom and the lab, but she had a great deal of difficulty when it came to visiting wards filled with sick human beings. And then, in her third year, she fell passionately in love with a young Bryn Mawr graduate who was permanently attached to another female medical student. The woman was responsive yet elusive. Instead of studying, Gertrude pined; she planned her life around seeing her beloved, or not seeing her. Sabotaged by the very parts of herself that she had been trying to cut off—the emotional, the uncontrolled—she failed her final exams and did not graduate. But by then, of course, she claimed she didn't care. She no longer wished to be a doctor. She had discovered that the only way to relieve her suffering was by writing. She was a novelist after all.

Virginia Woolf believed that no woman had succeeded in writing the truth of the experience of her own body—that women and language both would have to change considerably before anything like that could happen. She also believed that those who struggled toward the liberation of language—like herself and Stein and Eliot and Joyce—were bound to fail at least as often as they succeeded, and in this respect she judged Stein's "contortions" a generational misfortune. Woolf was eight years Stein's junior, and in many ways her experience and ambitions ran a parallel course: the death of her mother when she was thirteen; the influence of her dangerously overbearing father; the sexual "interference" inflicted by her stepbrothers; her love of women; her literary interest in male and female aspects of character and in an androgynous ideal. Also, like Stein, Woolf had enough money to write exactly as she pleased, and that was surely a major factor in allowing these two individuals, so different in their gifts and temperaments, to become the opposing poles of risk-all modern writing in a woman's voice.

The anguished novel Stein wrote when she joined her brother in Paris was an account of the love affair that had been tearing her apart. Stein's first full literary achievement, *Q.E.D.*—standing for *quod erat demonstrandum,* the conclusion of a geometric proof—was one of the few works she never tried to publish; although not explicitly erotic, it was plainly open about the fact that all three members of its sexual triangle were women. No explanations, no apologies, no wells of loneliness. Stein seems to have written the book as a kind of exorcism, incorporating in it actual letters and conversations, with the goal of restoring her serenity and never losing it again.

The book is fascinating not only for the information it provides about Stein but for the Jamesian acuity of its psychological portraits. It is hard to predict how far Stein might have gone in adapting this tradition—she was then twenty-nine—but the autobiographical narrator hints at why the attempt would be abandoned. She has always had a "puritanic horror" of passion, she confesses, and the pain of this love affair has rendered it absolute. "You meant to me a turgid and complex world," she rebukes her beloved—protesting also against the Jamesian tasks of emotional probing and dissection—before she heads off for a better or at least more soothing world of "obvious, superficial, clean simplicity." What she didn't know yet was how to find it.

After a year of frantic travelling, Gertrude moved in permanently with Leo in his apartment on the Rue de Fleurus. Because her erudite brother didn't approve of *Q.E.D.*, she wrote only late at night, hurrying to bed just before dawn so the birds would not keep her awake. During ordinary hours she was more than ever Leo's pupil, with everything to learn about the flourishing art of painting. When Leo took her to see the work of a young Spanish painter in a gallery owned by a former circus clown, she initially refused to chip in her share of the funds required to make a purchase. She thought the long figure of a nude woman had ugly, monkeylike legs and feet. The dealer offered to cut off the offending parts and sell just the head, but the strange Americans eventually returned and bought the whole thing. Picasso's *Woman with a Bouquet of Flowers* was soon followed into the Steins' salon by many other Picassos, as well as Matisses and Cézannes. But, most important, it was followed by Picasso himself, who thought Leo a dreadful bore but recognized in Gertrude a companion spirit. He had barely met her when he asked to paint her portrait. Uncharacteristically, he struggled with the subject mightily: after more than eighty sittings, he wiped out the naturalistic head and quit. He returned to the canvas only months later, and what he finally produced was not Gertrude Stein as she then appeared—no one thought the portrait looked like her—but the grandly masked *monstre sacré* she would become once she had forged her genius and had paid the price.

It was while she was sitting for her portrait, in the spring of 1906, that Stein thought out much of the book that first won her literary renown. Inspired by Flaubert's meticulously understated *A Simple Heart* (which Leo had set her to translate as an exercise in French), Stein's *Three*

Lives is a trio of stories about the grim existences of three women—two German-born servants and one poor black—who drift toward their fates in the airless atmosphere of a small American city. Stein's achievement was to make the writing seem as if shaped by the inner states of her characters: the childishly simple diction of those who had never willingly opened a book, the repetitiveness of those who had not much to think about or who were unused to being heard. The book was not published until 1909, and then in a tiny edition that Stein paid for herself. The reaction was astounding on almost any scale. "A very masterpiece of realism" was the general tenor of reviews. Many writers and critics began to see Stein's little book as the start of a truly American, unliterary literature, homely and vernacular and existentially unpresuming—a response suggesting that many writers and critics knew no more of the people she was supposedly writing about than Gertrude Stein did.

Like everything that she had written (but had not published) up until then, *Three Lives* is about Stein's autobiographical obsessions. Here are the vile but attractive father and the mother whose death is hardly noticed, the better-loved dead babies, and a continuous replaying of the tormented love affair, complete with psychological observations transposed from *Q.E.D.* But there is a change: the narrative voice is now so apathetic and the emotional temperature so low that it's no wonder the characters seem half-unconscious. The cause is not their social downtroddenness but the fact that *Three Lives* catches Stein in the very act of administering the emotional anesthesia that marked her style forever after.

The success of *Three Lives* coincided with two other important developments in Stein's life: Picasso's return from a summer in Spain with his first Cubist canvases, and the deepening of her relationship with Alice B. Toklas. The result was an explosion that shattered all her effortful old forms. The sudden outpouring of small, fragmented "portraits" and other glittering esoterica is usually said to have been inspired by—depending on one's source or one's inclination—either her desire to write like a Cubist or her need to conceal the erotic joy of her new attachment. Certainly, Cubism was vital to Stein, because it provided her with an intellectual rationale for doing exactly what she already urgently wanted and needed to do: keep her eyes on the surface. Facet it, mirror it, spin it around, and repeat it ad infinitum, but never go back underneath.

▪ ▪ ▪

When Gertrude first met Alice in the fall of 1907, she thought her the same type of "pure female" as her sister Bertha, and she was wary. "She listens, she is docile, stupid and she owns you," runs one notebook entry. A tiny, brittle woman, dressed in exotic fringed shawls that seemed to emphasize her ever-remarked-on Semitic features—"an awful Jewess, dressed in a window-curtain" was Mary Berenson's typical assessment—Alice also gave signs of possessing, according to Gertrude, "an exquisite and keen moral sensibility." And there was no doubt that she knew a genius when she met one. (Actually, she claimed that she heard a bell ring whenever she met one.) Furthermore, she was absolutely certain (now that she'd seen it) of the world in which she wanted to live. This was a very different world from the one she'd come from, back in San Francisco, where for the past ten years—since her mother's death, when she was nineteen—she'd borne the domestic burdens of a household made up of her father, her grandfather, and her younger brother. None of whom she, gladly, ever saw again.

Stein and Toklas were formally "married" in the summer of 1910, outside Florence, and that fall Alice moved into the Stein ménage on the Rue de Fleurus. And so it was that Alice B. Toklas "came to be happier than anybody else who was living then," as Stein wrote in *Ada,* a biographical portrait of her bride. Because Alice, for the first time since the death of her mother, had someone to tell her charming stories to: "some one who was loving was almost always listening." Listening and loving, loving and listening—the paired satisfactions now nearly replaced books and food. Notably, at this time Leo began to display symptoms of chronic deafness, and to starve himself in fasts that lasted as long as thirty days; he claimed to be writing a book on painting, but he couldn't produce a word. Just as Gertrude was becoming the man that being a genius required her to be ("I am very fond of yes sir," she wrote), Leo was turning into the very model of the mysteriously afflicted, hysterical woman she had once been so interested in trying to cure. But no longer.

With someone now listening to her so well, Gertrude began to pour out words without hesitation, revision, or second thoughts. Privately, there are the sweet-breathed burps and coos of utter infantile contentment: "Lifting belly fattily / Doesn't that astonish you / You did want me / Say it again / strawberry." The work she offered to the world made even fewer concessions to standards of sense and communication:

> One whom some were certainly following was one who was completely charming. One whom some were following was one who was charming. One whom some were following was one who was completely charming. One whom some were following was one who was certainly completely charming.

So runs the first paragraph of Stein's "portrait" of Picasso, which was published by Alfred Steiglitz (along with her "portrait" of Matisse) in *Camera Work* in 1912, a year before the Armory Show introduced modern art to *le tout* New York—with loans from the Stein collection—and made Gertrude Stein about as notorious as the painters whose outrageous principles she was said to share. ("The name of Gertrude Stein is better known in NY today than the name of God!" Mabel Dodge Luhan wrote to her ecstatically, and Stein replied, "Hurrah for *gloire*.") She didn't mind that most of her fame came in the form of parody and ridicule. "They always quote it," she pointed out in what may be her most truly modern observation, "and those they say they admire they do not quote."

All this was too much for Leo. He had hated Cubism from the start, and he didn't hang back from declaring Gertrude's work "Godalmighty rubbish." By late 1913 brother and sister were no longer speaking, and he soon moved out. The collection of paintings was divided more or less equitably, with Gertrude taking the Picassos and Leo the Renoirs; they split the Cézannes between them. (Picasso painted her an apple to make up for one of the Cézannes she lost.) Although Leo later tried to resume contact, she did not respond. "I have very bad headaches and I don't like to commit to paper that which makes me very unhappy," she wrote in one of her notebooks. More than thirty years remained of their lives, but Gertrude and Leo never spoke again. Alice claimed that Gertrude had simply forgotten all about him.

Stein had won enormous freedoms, but she chose to confine herself to a comfortably narrow space. In the midst of sexually explosive Paris, where the members of Natalie Barney's lesbian circle flaunted their glamour and their liaisons and their belly-dancing parties, Gertrude and Alice ran a salon that was a model of middle-class decorum. Despite occasional unorthodoxies of dress—Alice's window curtains, Gertrude's tents—and the Roman-emperor haircut Gertrude eventually got, they played the roles of two charmingly eccentric ladies who just happened to be man and wife (a fact as perfectly obvious to all as it was presumed unmentionable).

The effects of self-confinement on Stein's writing, however, were cruel. Beyond the occasional flash of wit or happy juxtaposition, her chains of words and repetitions come to suggest an animal's relentless pacing—cramped and dulled and slightly desperate. One feels that just outside the strict boundaries imposed by her pen lurked fathers and brothers and arguments and wars. In life, she could be head-on and courageous. During the First World War, she imported a Ford van and learned to drive it (except in reverse) to deliver supplies to hospitals all over France, and after the war she and Alice wrote to the many lonely American soldier "godsons" they had adopted. But nowhere in Stein's "literary" writing did she take in the experience: the wounded, the fear, the tenderness. She was writing less then anyway, giving over much of her time to advising and instructing (and sometimes to feeding and supporting) a group of young writers who found their way to her fabled door; in the twenties, as if by decree, the painters dispersed and the writers appeared. Among these, first in place and most fiercely devoted, was the twenty-three-year-old Ernest Hemingway, who sat at her feet and learned to write like a man.

"Gertrude Stein and me are just like brothers," Hemingway crowed to Sherwood Anderson in 1922. He brought her his stories to read and criticize, and they talked for hours while his wife, Hadley, was none too gracefully monopolized by Alice. (Holding off "wives of geniuses" was a stressful part of Alice's occupation; some had to be cornered behind large pieces of furniture.) He credited "Miss Stein" with getting him to give up newspaper reporting and to concentrate on his serious writing. He commended her method of analyzing places and people. And it was on the Rue de Fleurus that he was advised to go to Spain to see the bullfighting.

In 1925 reviewers of Hemingway's first volume of stories, *In Our Time,* recognized stylistic debts to Stein that were unmistakable: drastically short and unadorned sentences, repetition, a "naiveté of language" that suggested a complex, inarticulate emotional state. But for Hemingway the style clearly served as a kind of dam against an opposing sentimental pressure that was rarely if ever felt in Stein's prose. The new American hero was syntactically disengaged, because he'd been through hell and had already felt too much; his semiautism was part of his sexual equipment in a world in which physical courage was destiny and the only truly frightful things were women and emotion. Virginia Woolf—who used "virile" as an insult—particularly deplored Hemingway's exaggeration of male characteristics, for which she blamed the "sexual perturbations" of the times. (This was in 1927, and she was reviewing his aptly titled collection *Men Without Women.*) Woolf might have felt differently if she had traced the most famously virile of modern styles to its origins in the work of a woman—albeit a woman who liked to call herself "a roman and Julius Caesar and a bridge and a column and a pillar" (when all Virginia ever called Vita was "a lighthouse").

Are there male and female characteristics in writing? Male and female sentences? Is the comma a languishing feminine ruse, draining the strength of the tough male verb into a miasma of girlish uncertainties? Writing is self-exposure, and in the postsuffrage, neo-Freudian twenties the fear of what might get exposed was everywhere. Stein believed that the use of a comma was degrading and a sign of weakness because after all you ought to know yourself when you needed to take a breath. Woolf suggested that any woman who wrote in a terse, short-winded style was probably trying to write like a man. But at the heart of their difference is the fact that Woolf didn't see why a woman should want to do anything like a man; feminine generosity was life itself, and the necessary source of male achievement. One could hardly get further from Stein's perception of the sexes' division of properties. The two women met once—at a party in London, in 1926—and the revulsion was mutual. On Woolf's part, this was a matter of class and snobbery. "Jews swarmed," she wrote of the event in a letter to her sister. On Stein's part, there was defensiveness and bravado, surely based on a perception of the chilly atmosphere and perhaps, too, on the glaringly anomalous presence of a purely female genius.

By the late twenties, Stein and Hemingway had battled often, and they finally parted ways. She never spoke of what had come between them; he could hardly stop speaking of it. Sometimes he claimed that it was Alice, jealous of his relationship with Gertrude ("I always wanted to fuck her and she knew it"), who had caused the break. He also claimed that after Gertrude went through menopause she wanted no men around her except homosexuals—her "feathered friends," as he called all those whom he resented for usurping his place in the nest. Undeniably, there had been a change in the salon. Although Gertrude and Alice were as reserved as ever, they did live increasingly within a kind of tacit homosexual freemasonry. Indeed, it seems to have been this very reserve which was part of the attraction for such cautious old-world gentlemen as Frederick Ashton and Virgil Thomson and others from the highly sexually encoded worlds of music and theater and dance. In Stein's new role as a professional collaborator writing opera and ballet librettos, and also as a friend, she seems to have developed the appeal of a more or less inverted Mae West: a good-humored woman in male drag, a warm and wise mama who not only took in all her gay sons but was gay herself.

None of which was known to the public, of course, when *The Autobiography of Alice B. Toklas,* written in 1932, suddenly turned the world's most famously obscure writer into the best-selling author of a Literary Guild selection. From the start, Stein had thought of the *Autobiography*—actually it is her own biography, written from the dazzled point of view of her companion—as an embarrassingly traditional, moneymaking venture, on a different level from the meaningfully experimental work she still composed at night. In fact, this magical book doesn't resemble anything else in Stein's work—or, for that matter, anything else in American literature. Its only models seem to be those other famed "Alice" books, by Lewis Carroll, and the social surrealism of Oscar Wilde, which further muddles the question of the sexual significance of a writer's style.

The book is a modern fairy tale: the story of a golden age of art in Paris, when geniuses regularly came to dinner and were cleverly seated opposite their own paintings so that everyone was made especially happy, although everyone was happy anyway—this was back before death and divorce and success—and Henri Rousseau played the violin and Marie Laurencin sang, and Frédéric of the Lapin Agile wandered in and out with his donkey. And into the middle of this wonderland walks the sensible American Alice B. At her very first dinner, she finds herself next to Picasso, who gravely asks her to tell him whether she thinks he really does look like her President Lincoln. "I had thought a good many things that evening," she reports quite as gravely, "but I had not thought that." Ever temperate, Alice is the quiet but quizzical eye at the heart of the storm. Her equanimity, like her pleasure, is absolute. The book indulges in some small revenge—Leo is not mentioned by name, Hemingway is imputed to be a coward—and it leaves us with a sense of loss that is as profound as it is muted. Once, there was so much life all around that one had to hurry to bed before dawn to have any hope of sleep: "There were birds in many trees behind high walls in those days, now there are fewer."

With the book's success, Stein fell into a profound depression. Even her new *gloire* didn't help, at first; she was fifty-nine, she was having her first major success, and it was for the wrong thing. In *Four in America,* a particularly reader-resistant work she completed about this time, she rose—briefly, rather thrillingly—to a clear explanation of her intended goals and values:

> Now listen! Can't you see that when the language was new—as it was with Chaucer and Homer—the poet could use the name of a thing and the thing was really there? He could say "O moon," "O sea," "O love," and the moon and the sea and love were really there. And can't you see that after hundreds of years had gone by and thousands of poems had been written, he could call on those words and find that they were just wornout literary words? . . .
>
> Now listen! I'm no fool. I know that in daily life we don't go around saying 'is a . . . is a . . . is a . . .' Yes, I'm no fool; but I think that in that line the rose is red for the first time in English poetry for a hundred years.

Alas for good intentions. When Stein returned to America for the first time in thirty years, in 1934, to attend a performance of Virgil Thomson's opera (to her libretto) *Four Saints in Three Acts,* she also gave a series of lectures around the country. Directly exhorting her audience, Stein made headlines (MISS STEIN SPEAKS TO BEWILDERED 500) but clearly failed to win the understanding she was after. The work of the woman who claimed to believe so fervently in the immediacy of language and the recovery of meaning remained synonymous with obscurity and confusion. (In the 1935 movie *Top Hat,* Ginger Rogers giggled that an indecipherable telegram "sounds like Gertrude Stein.") Any attempt to assess Stein's literary achievement raises many old questions—not only about her judgment and credibility but about the relationship of theory to art in the troubled history of modernism.

When Stein and Toklas returned to Paris in the spring of 1935, Toklas took it upon herself to ship copies of Stein's manuscripts back to America for safekeeping. This was the full extent of their preparations for the possibility of hard times ahead in Europe. Stein couldn't believe that anything would really happen, and certainly not to them—two elderly Jewish ladies with a bit of fame and no political interests. Their political indifference was now dangerously compounded by Stein's long-standing way of handling all serious unpleasantness: pretend it isn't there, and then tumble into nonsense or baby talk, so that perhaps it will be persuaded that you are not there, either—or, at least, that you are not a reasonable target. In her early Paris days, she had written to a friend who had apparently mentioned the Russian pogroms against the Jews, "The Russians is very bad people and the Czar a very bad man." In the mid-thirties, her thoughts about Hitler were hardly more sophisticated. She and Alice were staying in the countryside, at Bilignin, in 1939, when war broke out. They hurried to Paris in order to close up their apartment. Then, taking one Cézanne and Gertrude's Picasso portrait, they returned—against all advice—to Bilignin, which soon fell under the jurisidiction of Vichy. And there they spent the war.

How did they survive? Largely, it seems, through a French admirer and friend who was appointed head of the Bibliothèque Nationale under the Occupation and issued several requests that they not be disturbed. They lived quietly and scrounged for food; they sold the Cézanne and liked to tell visitors surprised at the quality of their dinner that they were eating it. They watched the German army march into the land and, a long and burning time later, straggle out; German soldiers were billeted for a time in their house, as were American GIs when at last they arrived. This experience is recounted clearly and movingly in a book Stein completed in 1944, entitled *Wars I Have Seen.* Sadly, the book is out of print, and its failure to meet a modernist criterion has kept it from inclusion in the Library of America compilation. Yet it is one of the few of Stein's works that might be called essential, because it tells the ending of the fairy tale.

There is much here that is richly, lovingly observed, about the women's daily life and that of their neighbors and about French pragmatism and courage. Typically, whatever could not be lovingly observed is passed over—or nearly so. For now history almost catches up with Gertrude Stein, and forces her into confrontation. One feels her struggling with the effort not to look away: the very term "collabo"—which is all she manages to spit out of it—causes her to stumble on the page, falling into a repetitive stutter that seems not a mannerism but a kind of seizure. Her attempt to address anti-Semitism begins with an early memory of the Dreyfus Affair, but she hasn't advanced beyond a sentence when she segues into senseless babble: "He can read acasias, hands and faces. Acasias are for the goat. . . ." "Acasia" is not even a word (and Stein never made up words; she thought such arrogant idiocy to be the province of Joyce). "Acacia" is the name of a tree, but "aphasia" means the loss of the ability to speak. It is as though Stein were making her own diagnosis, or as though a part of her reason were watching the rest of her mind run away.

After the war, back in Paris, the famous salon was filled with GIs eating Alice's chocolate ice cream. Gertrude wrote down what they said and how they said it as though they were the new poets of the age. She adored them, she celebrated them: they were, after all, young men and heroes. But something in her attitude was changing. Along with the ice cream, she dispensed correction; she worried that the great liberators were taken in by the flattery and politeness of the postwar Germans, or that they saw the world in terms of movies. By and large, it seems to have dawned on her that there was a lot these callow demigods ought to learn before the world was put in the hands of men, even the best and noblest men, ever again.

Although she was over seventy and weak with cancer, Stein was very eager to work. In late 1945 she began a second opera project with her old friend Virgil Thomson. It was his idea that the setting be the American nineteenth century; it was her idea that the hero be a heroine, the suffragist Susan B. Anthony. Stein completed the libretto for *The Mother of Us All* just before her death, in July 1946. Although the work is predictably baffling, Stein maintains an exceptionally strong dramatic focus on the character of Susan B. She had done a considerable amount of historical research—shocking in itself, given her usual methods—and it is Anthony's public concerns that dominate the text: the disparities between the sexes, and her passionate conviction that women are stronger and must lead.

For men are afraid, Stein's heroine observes: "They fear women, they fear each other, they fear their neighbor, they fear other countries and then they hearten themselves in their fear by crowding together and following each other, and when they crowd together and follow each other they are brutes, like animals who stampede." As for women, they are afraid not for themselves but only for their children: "that is the real difference between men and women." Stein shows Susan B. at the end of her life, when she knows that all her work has failed. She has helped to win the vote for black men, but she will die before it is granted to women, white or black. When someone attempts to comfort her by saying that women will vote someday, her despair only deepens: she dreads that when women have the vote they, too, will become afraid—that they will become like men.

How astonishing that Stein can now voice her lifelong wish—"women will become like men"—as a dreaded possibility. A new sense of the value of what women had traditionally done and been may even have given her succor in these last months, as she looked back on her own life as the mother of so much and so many. Perhaps by then she had realized that in her years of giving and feeding and advising and encouraging and (is there another word for it?) mothering—in the continual dispersal of herself to men whom she loved and admired, to geniuses and soldiers and cowards alike—she had inadvertently lived the life of the most profoundly womanly of women, and that it had been good.

Sex, Lies, and Thirty-five Thousand Pages

Anaïs Nin

"Have I been less brutal, less passionate than you expected? Did my writing perhaps lead you to expect more?" The lover intimidated and somewhat abashed by the potency of his own writing is the forty-year-old Henry Miller, penniless and unpublished and bumming around Paris, and the work in question was his explosive new novel—"the Paris book," as he'd planned it, "first person, uncensored, formless—fuck everything!"—which he would title *Tropic of Cancer*. In the late fall of 1931, dreading a rainy and rheumatic winter and with no roof over his head, Miller had sent pages of his unfinished manuscript to the wife of a well-to-do American banker, a woman known for her literary aspirations, in hopes of a few golden eggs. The following March, he found himself laying the goose.

Anaïs Nin Guiler bore no resemblance to the conventional image of a banker's wife. Indeed, she cultivated a highly successful disguise for this embarrassingly bourgeois condition, draping herself in gilt shawls and flowing capes and the exotic, gypsifying ornaments her husband's profession afforded her. Nin's celebrated *Diary,* eventually published in seven volumes drastically edited and expurgated by the author herself—six between 1966 and her death in 1977, and the seventh in 1980—offered a widely appealing portrait of an independent woman in an artistic milieu. Few readers of this extensive life can have suspected that it was largely out of her clothing and travel allowances that Nin was able to rent and furnish one of the best-known corners of 1930s bohemia, the small apartment in Clichy where Henry Miller lived the quiet days that allowed him to finish his book.

Before the *Diary* made its public appearance, Nin was known, if known at all, as the author of several strenuously avant-garde novels and stories of narrowly female sensibility. Works like *Under a Glass Bell* and *Children of the Albatross* were printed on her own press, in New York, in the forties, or, when commercially published, were bought up by the author in dispiriting quantities at Womrath's for forty-nine cents a copy. But the publication of the *Diary* reversed Nin's literary fortunes and brought her international renown—she won France's Prix Sévigné for autobiography in 1971—and, in the United States, a position that might be characterized as the forefront of the rearguard of the women's movement. The suggestive glamour offered in these volumes—a suburban Paris house decorated à la Scheherazade, a shifting cast of uncertainly affiliated men—gave way in the late 1970s to the franker allure of Nin's newly published "erotica," two volumes of stories that had been written in the forties for a private client and glimpsed before only in the *Diary.* These collections, approved by Nin but published after her death, became best-sellers and the cornerstone of a new genre of frankly titillating fiction (a.k.a. "the new sensual writing") by and for an unashamed new audience of women.

The astounding commercial success of all these works has led to the publication, potentially unending, of a second series of diary volumes, consisting of sections expurgated from the first published version, with a few passages reprinted to provide a narrative thread. Titled *A Journal of Love,* for purposes of distinction, this new version of the life made its debut with *Henry and June*—the subject of a 1990 movie—and followed, in 1992, with the straightforwardly titled *Incest.* Decade by decade, Nin—or her reputation—has been reinvented. The arc of her career is traced with a suitable tabloid grandeur in the title of a 1977 article about her in a Japanese magazine: "The Incredible Contents of Anaïs Nin's Pornography: Queen of American Avant-Garde Literature from Nouveau Roman to Jazz: Sex Hunter Who Taught Henry Miller How to Love."

■ ■ ■

In 1931 the banker's wife was twenty-eight. She had been preparing for Miller's arrival for a long while—if not quite from the moment in adolescence when she envisioned her life bound to "a great writer and I would help him a tiny little bit in the fantastic, poetic, imaginary chapters" (a fantasy immediately checked by the pragmatic "Perhaps I would do better to write the book myself, and he could correct it"), then surely from the crushing sexual and intellectual disappointments that immediately followed her marriage. Nin's worshipful young husband, Hugh Guiler, had offered everything but the "support and leadership" she craved. "Where shall this lead us?" she begged of her diary in 1923, as a twenty-year-old bride of little more than a month. "I, who believed myself made to cling, thrown upon my own strength." She had hoped, she added with some poignancy, that the fulfillment of love would end the compulsion of the diary—already a decade old. But the diary is, of course, where she confided this hope, and the pages went on accumulating, and the same stories kept appearing: Anaïs is beautiful, she is admired, she experiences many passionate but inconclusive flirtations, she can't sleep and she cries when alone and she feels "an almost physical pain in my body, an ache, a hunger, an emptiness and burning which nothing soothes," as she wrote in 1928. "Music, men's admiring eyes keep me stirred like the earth in the fields." By the time Henry Miller arrived at her door, she was up to volume thirty-one.

To judge by their own accounts, the woman who would be acclaimed as the Eternal Feminine descended on the man who staked out the Irremediable Masculine like—to borrow a phrase from Nin's sometime friend Rebecca West—a fold on a wolf. Daunted by Nin's position, by her house and her clothes, even by her sweetly accented English—she was born a Cuban citizen in Paris and grew up in France and Spain and, for many years, Queens, New York—Miller felt obliged to keep a respectful distance until the day she sat beside him in the café Chez les Vikings, on the Rue Vavin, and, eyes demurely lowered to the page, read to him from her journal about the effects his writing had upon her. What appeared to be an impulsive revelation of a young woman's bursting heart was actually a rather well-practiced strategy. Nin's *Early Diary*—the official title of four volumes brought out after her death, which take us from 1914 to the point in 1931 at which she had begun the published story herself—were never cut or emended by her, and offer several demonstrations of the efficacy of selective journal recital on a fair range of male targets prior to Miller. For

Nin, this had become a prime method of getting around the limits imposed by her china-doll looks: "Curse my eyes which are sad, and deep," she lamented in a passage that she did process for publication, "and my hands which are delicate, and my walk which is a glide, my voice which is a whisper, all that can be used for a poem, and too fragile to be raped, violated, used." The famous diary—famous in her circle even in the 1930s—could be as much a tool of seduction as a record of it, employed to spur the appropriate aggression while betraying none in the recording angel herself.

But Miller's nervous question remains: was he brutal enough, passionate enough, was he a match for her literary expectations? Was anyone? In the diaries dating from her early married years, Nin liked to compare herself with Madame Bovary, although she insisted on an important distinction: "Unlike Madame Bovary I'm not going to take poison." The heroine was taking charge of the story, and it wasn't going to end up as tragedy. But by the winter she met Miller, a new and far more encouraging model of literary adultery had come to her attention. Nin had begun reading D. H. Lawrence in 1929, and her first slim book, published by an English firm in Paris in 1932, was an "appreciation" of his work. Nin's self-appointed task was to defend Lawrence's ideas about women from the charge of being "antediluvian"; she argued that he dealt with timeless "quintessences," such as the fact that (in her words) "the core of the woman is her relation to man." She declared the banned and infamous *Lady Chatterley's Lover* to be "his best novel" and, moreover, "our only complete modern love story." Against the gentle and ingenuously American warning that Hugh Guiler offered when his wife intimated her hunger for "experience" (they were free, he told her, without religion or convention to bind them, and yet "you have seen in literature and in France what your ideas, carried to extreme, lead to: mediocre unfaithfulness"), Lawrence's ideal of sex as transcendent rite offered a defense and a high justification.

To compound matters, about a year after her discovery of Lawrence she began to read Freud, and became convinced that she was suffering the nervous symptoms of hysteria, and that the cause, as Freud proposed, was in sexual repression. For Nin, Freud gave the imprimatur of science to the same quest for physical fulfillment that Lawrence cast in terms of religion. Imagine Emma Bovary reading Lawrence and Freud: how was a woman—or such a woman—to think that she wanted anything else?

"But Lawrence men are so rare, so rare," Nin's cousin Eduardo sighed along with her in 1930. "Have you really met one?" To which Nin sadly replied, "Pieces of Lawrence, just pieces." She was quick to recognize the whole man, or as near to it as the world could offer, in Henry Miller. "I began by adoring Lawrence, and I end by worshiping a man so much like Lawrence, like Mellors," she wrote, equating the writer with his sexual hero—the Heathcliff of the circa 1930 liberated woman. But Nin had good reason to confound Henry Miller the writer with the hyperbolically exuberant stud who narrates his own audacious novel *Tropic of Cancer* ("After me you can take on stallions, bulls, rams, drakes, St. Bernards"), since that narrator is also named Henry Miller, and the author himself couldn't much tell them apart.

Nin's earliest full report on her research into life with the Lawrentian male (published in *Henry and June*) followed a visit to Miller's shabby hotel room in March 1932 and was, in fact, all that Lawrence or Freud could have wished. "I cannot conceal it. I am a woman. A man has made me submit," Nin enthused. "Oh, the joy when a woman finds a man she can submit to, the joy of her femaleness expanding in strong arms." Miller, in his letters, seems a good deal less certain, at the start, of who had submitted to whom, but his tone changes rapidly to the more deliberately winning likes of "And Tuesday prepare to be raped" (a promise accompanied by a characteristic plea for funds: "There isn't a bit of food left"). But just eight months after the first momentous submission, Nin was railing bitterly against "the weak man whose weakness kills me," a torment inflicted upon her now by none other than Henry Miller: "I did everything to find a leader! And again I am cheated." Although she believed that he had been sexually mistaught by an overaggressive wife, the infamously flagrant June Mansfield (from whom Nin was even then in the process of detaching him), she had yet "counted on Henry becoming a man when confronted with a real woman—a really *passive* female. And he is baffled—baffled by my submission."

Miller, with just a bit of diary prompting, quickly made up for his lapses, and reappears a page later quite satisfactorily unbuttoning her dress. But this—or he—was not enough. Nin's vision of Miller's genius, and her financial support, were constant through the mid-thirties: she paid the rent, provided food and nattier clothes than any he'd worn before, and sent him money for a whore as a token of her love and her bohemian open-mindedness. Most important, she backed the printing of his book. But by the time *Tropic of Cancer* was going to press, in 1934, it was another lover—the psychiatrist Otto Rank—who put up the five thousand francs that the publisher required.

Otto Rank was not the first lover Nin took after the plunge into her affair with Miller; he was not even the first of her psychiatrists she took.

Incest, the second volume of the unexpurgated edition of the diary, reveals that she was having affairs with, basically, all the men who had been deemed worthy of mention back in the expurgated volumes, excepting Antonin Artaud, whose impotence proved something of a relief, given the laudanum stains around his mouth. And, as the title suggests, the list of lovers included her father. In fact, the only noncandidate in Nin's search for a Lawrentian hero during the thirties appears to have been her hard-pressed husband, who was then undergoing analysis for the cure of such ills as, in his wife's words, "his over-devotion to the bank," "his feminine fear of bullies," and "his unresponsiveness to my work." It is something of a triumph over biology that Nin's incestuous relationship with her father, a Spanish musician and dandy who had disappeared from her childhood decades before, is far less disturbing to contemplate than her affair with the psychiatrist (Rank's precedessor, Dr. René Allendy) who was administering to her husband's psyche. Not even Charles Bovary underwent such an indignity.

Still, Dr. Allendy did work at "curing" Hugh Guiler—that is, at making him "less dependent" on his wife, since she had declared that only then would she feel free to leave Guiler for Allendy. (Rank doubtless agreed to finance Miller's book for the same sort of reason. Nin stuck by her methods.) We are brought close to farce—if only there were some wit involved—in a scene where Nin's husband, psychoanalytically inspired, manages at last to ravish his wife (*"And I had always wanted my dress torn!"*), who, barely out of his arms, plans a consolatory visit to the psychiatrist to whom her husband will soon be crowing about his triumphant virility. But Hugh Guiler's triumph came too late, and Allendy eventually lost out altogether, passing in Nin's account from a "giant" with "idealistic fingertips" to a man of "pulpous flesh" and, almost as a matter of course, "sexual timidity." Her search went on.

The real and bottomless subject of Nin's diary is not sex, or the flowering of womanhood, but deceit. And the lies to the men in her life are only the beginning. These are the deceptions of which she was proudly conscious, the crafted duplicities and quadruplicities in which she snared her sense of "livingness," and which on her better days she laughed over—like the clever demimondaine she sometimes seems to be—and on her sorrier days she rationalized: "I only lie as doctors do," she told Allendy, "for the good of their patients." Was there a deeper morality to consider? "I bring home to Hugo"—as she called her husband—"a whole woman, freed of all 'possessed' fevers, cured of the poison of restlessness and curiosity which used to threaten our marriage," she wrote in March 1932. "Amorality, or a more complicated morality, aims at the ultimate loyalty and overlooks the immediate and literal one."

Nin's volumes are rife with lies large and small, from passages of actressy virtue which can have been intended only for her husband to read—these are in the *Early Diary,* which she didn't edit, but which she did occasionally leave artlessly open on her desk—to the general omission of her means of material support from the volumes that she did edit, to the saintly gloss on her own character ("I palliate the suffering of others"), and to such niggling details as whether or not she'd met André Gide. (She reported in volume one that she had, but she hadn't, thus taking in her readers as she had once happily taken in her friends. "I don't believe she ever met Gide" was June Mansfield's rather astute exit line from the hopelessly literary triangle that her marriage had become.) Such lies had the clear purpose of protecting Nin's private advantages, and of helping to set an image for the public gaze that she always anticipated. Countless other falsehoods seem to have been useless, told simply "to make life more interesting. To imitate literature." And then there is the biggest lie of all, the one she used to deceive herself and to excuse herself: that she was an artist, and so a being of a separate order.

The only thing worse than having an artistic temperament, Henry Miller once groused, is thinking that you have one. Nin recalled herself as a girl of eleven, in 1914, arriving in New York with her mother and her brothers on a ship from Barcelona and descending the gangplank with her brother's violin case in her arms. She couldn't play the violin, but she wanted her assembled relatives to know that an artist had arrived. On board ship she had begun a diary, as an album of the trip, complete with drawings and pasted-in postcards; she continued it in her new home—changing from French to English in her teens—as a retreat for her thoughts and for extensive considerations of her appearance and character. (At fourteen she listed "lies" and "vanity" on her debit side, "sacrifice" and "charity" to her credit.)

In a home where everyone else had musical talent, the diary also became a demonstration of her artistic nature. The notebooks were dutifully bound in leather by her mother, perhaps in deference to the numerous letters to "Cher Papa" that they contained, all carefully copied out before being sent. It had taken the girl about a year to realize that her father, whom she had not seen for some time even before leaving Europe, was kept from rejoining his family not by the temporary disruptions of war but by the far more significant phenomenon, as it affected her, of a change of heart. Nin later explained her enormous twin appetites for artistic acclaim and male admiration as results of her father's desertion, much the standard and reasonable psychological assumption. (The explanation originated in analysis, as did a number of her compensating affairs.) But even as an undeserted child of seven she signed her stories "Anaïs Nin, Member of the French Academy," and seemed well on the path to her teenage goal of being "Loved and Published."

At twenty she was trying her hand at novels and plays, and writing in her diary of her friends' reactions. "Make your people walk in a real world," these first critics begged her. "Give us more concrete, pictorial writing." Since she had already recognized that such accomplishments were beyond her, she redirected her energy toward "outer forms," like sewing and the decoration of her house. But when the National City Bank transferred her husband to its Paris office in 1925, Nin discovered Surrealism; and in magazines like Eugène Jolas's *transition* she found models she could imitate and standards she could meet—or, at least, standards that few could say for sure she wasn't meeting. ("We are not troubled by manuscripts we do not understand," Jolas wrote, and his intrepid magazine, begun in 1927 and specializing in "the modern spirit," went on to gain distinction with eighteen segments of *Finnegans Wake* and with works by Samuel Beckett, Gertrude Stein, William Carlos Williams, and Gide himself.) And so Nin followed an attempt at a straightforward novel, which she'd called *Aline's Choice,* with a pair of "prose poems" called "Winter of Artifice" and "House of Incest," both composed in a new voice that she referred to as her "Sybilline tongue." (Sample sentence: "I carry white sponges of knowledge on strings of nerves.")

The "modern spirit" became an excuse for unintelligible writing, as her father's desertion was the excuse for her games of betrayal. In the summer of 1932, she handed Miller a sheaf of thirty new pages and he confessed himself mystified. He told her, in exasperation, that the story would have to be read a hundred times to be understood. Nin was discouraged for a moment. "But then I thought of *Ulysses* and the studies which accompany it." Pressing further the following year, Miller wrote her several letters about her writing, and about "what I can do to save you," which make even his wonderfully sexy, hungry letters about her body ("I love your loins, the golden pallor, the slope of your buttocks, the warmth inside you, the juices of you") seem routine expressions of devotion.

Miller had come a long way from his early novels, written in a forced attempt at realist conventions; both *Moloch* and *Crazy Cock* (published for the first time only in the 1990s) show the distance that he vaulted into the open air with *Tropic of Cancer.* And he saw no reason yet why Nin, too, could not learn to conquer what seemed her limits. Of her new novel, *Alraune,* he wrote her in the spring of 1933, "When you come again, and every time you come, until you get blue in the face, I'm going to trot it out and we're going to work over it slowly, patiently, lovingly, scathingly. . . . It must not die, in the way things do, through oversight, laziness, neglect." Her chosen style—she now had in mind Rimbaud, Laforgue—was exceptionally difficult. "It is either flawless, or else it is gibberish," he wrote. "I mean there are no intermediate degrees. A rocket, a shooting star—or else cinders." She had the genius for it, he assured her. "But you have no language in which to express it. And that can be tragic, if it is not overcome." He made a number of specific suggestions—toward concreteness, toward reality—and he rewrote, and encouraged, and finally he begged her to give up the diary. "It is good only if you recognize it for what it's worth—otherwise it is dangerous, poisonous, inclined to make one lazy, facile, self-contented. . . . I'm going to help you. Jesus, believe me. Ask me, make me!"

In response, Nin passed from initial acceptance and a resolve to work harder—"I am saved," she teased that April—to a fury of rejection a few months later, brought on by the further insult of a literary agent having rejected both a section of her diary, which she now openly hoped to publish, and her new novel, which he dismissed as "overdrawn, overwritten, overintense" and, despite its avowedly modern spirit, containing pages suitable to 1840. This was too much; the diary, her pride, appeared to have indeed been a waste of time—or, worse, a kind of disease, sapping her strength for the real tasks of her life. Not that there was much strength left, even now, for those tasks Miller tried to impose. It was already "getting more and more difficult to make four men happy," she wrote, and that summer she added her father to the list. ("Let your father devour you," Miller advised, although apparently unaware of the extent of the relationship; "it will give him dyspepsia. He has no idea what he's trying to bite off.")

She had come to imagine that this unnatural affair might mean the death of the diary, as she had once thought that her marriage would—this time, through the ultimate redress of its oldest, motivating hurt. Instead, as had happened after her marriage, she seemed to require its comforts all the more. If Nin often saw herself as a saint—"If all of us acted in unison as I act individually," she wrote, "there would be no wars and no poverty"—the diary was her attribute, a martyrdom she carried with her everywhere.

She was, in fact, holding the latest volume in her arms—one can't help recalling the child with the violin case—when she went to see Otto Rank in the fall of 1933. As a young man in Vienna, Rank had been part of Freud's inner circle, very nearly his adopted son, but he had broken with the Master over central tenets, from the Oedipus complex to the advisable length of analysis, and he was famous now as an apostate—Miller admired him as an "outlaw"—and as a specialist in the psychology of the artist. Rank, too, wanted to break her diary habit, to stop her from "nurturing the neurotic plant," as she put it. His first request was that she leave with him the volume she carried. She was appalled: it contained all the lies she'd planned to "interest him" with, worked up while she was waiting for her appointment. And yet, feeling mastered, she was also thrilled.

As the invincible diary itself reveals, their affair began the next spring—she chose the day the way she chose her hyacinth-blue dress for the occasion—when she was three months pregnant with Henry Miller's child. She disposed of that problem in August 1934, in a gruelling abortion, revealed only with the 1992 publication of *Incest.* She was entertaining the idea of becoming an analyst ("I might as well make a profession of a hobby") and studied briefly with Rank that summer before the classes began to bore her. She followed him to New York when he decided that his profession required a greater supply of neurosis than could be located among the French, whose acceptance of "the separation between love and passion," as he told her, gave them access to both, without need of professional intervention.

In Manhattan, Rank's fortunes picked up considerably. Nin, arriving in late 1934, appears to have worked as his general assistant, smoothing translations of scholarly papers and even seeing patients on her own. Almost immediately, she sent a generous sum of money to Miller, who had just discovered that she was not travelling on bank business with her husband. "What the hell *are* you doing?" he fired back. "In four days one can't really earn all that dough! Unless you're getting the rates of an analyst."

In a panic of jealousy, Miller sailed to New York (using her money—or Rank's—for the ticket), and while keeping generally out of Rank's way, he actually conducted sessions with several patients himself. (Years later, he revealed his method: "All you have to do is listen and speak only when necessary and then in a soothing voice.") But by spring both Nin and Miller were eager to get back to Paris. She'd found she could not earn a living as an analyst without Rank's backing, and both the man and the work had grown too demanding. ("I knew the time had come to leave New York or I would be consumed in service and healing.")

The publication of *Tropic of Cancer,* in Paris in September 1934, had made Miller an established writer; and in Europe, at least, he was now approaching a kind of celebrity. A second edition of the book appeared in 1935, bearing endorsements from the most elevated of English literary voices: Ezra Pound, Aldous Huxley, and even T. S. Eliot, who called it "a rather magnificent piece of work." (Eliot, who had denounced Lawrence as a heretic only a year before, must have seen in Miller the bleak confirmation of his view of the horror of sex in an age devoid of grace. These two American refugees, drawn to the antipodes of London and Paris, divide the divided modern man neatly between them.) *Tropic of Cancer* was, of course, unpublishable in the United States and in England, but its reputation was felt, like an underground explosion, even where the book could not be seen. During the war, it became an overseas GI classic, and for nearly three decades before the novel could be legally imported into the United States, literary Americans in Paris considered themselves nearly duty-bound to smuggle a copy home. It was in this way that a large number of readers first heard of Anaïs Nin, for the preface that she and Miller had worked on together appeared in her name.

In fact, by the late 1930s Nin was becoming estranged from Miller. His new writing seemed to her too crude, while her own was everywhere still rejected; and in any case, she had found another man and another scenario to play. Her new lover was South American—an "untamed Inca from Peru, too tall, too wild," or, simply, "Heathcliff"—and Nin's diary now displays her removing her nail polish and stashing her jewelry away in order to attend meetings about Republican Spain. She addresses envelopes for the cause and looks down on the rich at Maxim's, and also on Miller's determined political indifference. (Miller, who espoused the pacifism of Taoist philosophy, did give his corduroy jacket to George Orwell when the English *antifascista* passed through Paris on his way to Spain; but as a friend of Miller's pointed out, he would have done that even if Orwell had been off to fight for the other side.)

There was no replacement for Nin in Miller's life. He was intent on retaining his claim to her, however inexclusive, and however inexplicably compounded of greed, lust, gratitude, resentment, and love. He couldn't save her as a writer—that ambition had proved a disaster—but, what might please her more, he could try to sell her as one. He had given up on her fiction, but in his Surrealist-tinged enthusiasm for nonclassical forms, for the work of the amateur or the child (or even, to press a point, of the mad), he found a way to champion the diary. His campaign began with a fittingly Surrealist notice in a Paris golf-club magazine run by Americans, called *The Booster*, in 1937, announcing "the publication in January 1938 of the first volume of Anaïs Nin's great diary," a "faithful reproduction, in a child's French," to be issued in a limited edition of two hundred and fifty signed copies, by subscription only, with checks to be sent to Henry Miller. "In the event of a world war or a universal collapse of world currencies all monies received will be refunded at par."

It was *The Booster* that suffered a collapse, however, and there were no sales. But Miller had taken it upon himself to write a critical study of the diary, and that appeared, albeit without a subscription form, in the more stable context of T. S. Eliot's magazine, *The Criterion*, in late 1937. Windy and wildly foolish and occasionally brilliant, Miller's essay opens with the assertion that Nin's "monumental confession" would one day take its place "beside the revelations of St. Augustine, Petronius, Abelard, Rousseau, Proust," and more or less ascends from there. Nin thought the essay, called "Un Être Étoilique" (she'd invented the term *étoilique*, as a play on *lunatique*, to describe her star-touched self), merely a demonstration of Henry's usual fantasy, having more to do with the workings of his mind than with her achievement. While there is obvious merit in this view—Miller's flights of prose hardly require a subject to keep them aloft—there are also passages where he swoops suddenly into a tight and deadly accuracy:

> The illusion of submergence, of darkness and stagnation, is brought about by the ceaseless observation and analysis which goes on in the pages of the diary. The hatches are down, the sky shut out. Everything—nature, human beings, events, relationships—is brought below to be dissected and digested. It is a devouring process in which the ego becomes a stupendous red maw.

But the observation that was calculated to incite interest—and the one that has characterized assessments of Nin ever since—had to do with her sex. Of the later diary volumes Miller wrote, "It is the first female writing I have ever seen: it rearranges the world in terms of female honesty." Bitter as that judgment is, given what Miller knew about Nin's honesty, its implications were exciting, rather like Alfred Stieglitz's legendary "At last, a woman on paper!" when he first saw the "private" sketches of Georgia O'Keeffe. Miller's assessment was based on the diary's frank descriptions of sexual experiences (often involving Miller himself), but it is difficult to make that out from the evasive terms of his discussion, which centers on style rather than on content. Nin's "female writing," by his account, consists of "larval thoughts not yet divorced from their dream content," thoughts "never once arrested so as to be grasped by the mind." The diary offers "the opium world of woman's physiological being," with "not an ounce of man-made culture in it; everything related to the head is cut off." All that he had objected to before, all that he had tried to save her from—the secondhand surrealisms and the self-conscious "femininities"—are no longer presented as destructive intellectual flaws but as a biological absolute. The trick, then rather novel, lay in the elimination of the standard, its declared nonapplicability. Nin wasn't a failed writer, or a lazy writer: she was a woman writer.

Miller's essay was included in his first collection to be published in the United States, called *The Cosmological Eye.* This was also his first book to appear here legally—in 1939, the year that *Tropic of Capricorn* (subtitled *On the Ovarian Trolley*) joined *Tropic of Cancer* on the shelves, or, rather, behind the counters, of such rare bastions of literary freedom as New York's Gotham Book Mart. The Gotham was Miller's first stop on his return to America, in January 1940. Nin and her husband had come back the year before and had taken an apartment in Greenwich Village. There she held on to a small reputation—an indistinct sense that she was or had been someone—while writing fiction that still had no apparent audience.

In her *Diary* for these years Nin cries great poverty, and she did take a job writing pornographic stories at a dollar a page, but this seems to have been little more than another round of the bohemian game. Her husband, who is dropped entirely from the published account at this point (in earlier volumes he appeared as "the Humorous Banker" or "the Little Magnate"), had been made an assistant vice president of the bank. She may have longed for the independence of means with which to leave him—"I am a prisoner of material necessity" was how she had put it back in the days with Rank—but she was going nowhere without it, and at a dollar a page it would take some time.

Through the early forties, Nin still occasionally sent money to Miller as he ran out of gas or food while travelling across America, making notes for *The Air-Conditioned Nightmare.* When she mentioned at one point that she was thinking of going out to look for a job—in the tone of someone going off-stage to clean a gun—Miller replied, "I beg you not to. If anybody is to take a job it's me, not you. . . . But you have a husband to look after you." He had already begged her several times to "make a break," to join him: "The only bugaboo I see is poverty and that is no longer a bugaboo to me. . . . You made me free, liberated—now free yourself. I have no attachments. And no concubines either. Please believe in yourself, in your own powers." But she was furious with him for staying away so long, and by the time he got to California, where he found a home, the rift was emotional and philosophical, and irreparable. "Why you persist in thinking I am blind or indifferent to your suffering I don't understand," he wrote her in 1943, "nor am I going to argue with you about the relative merits of Stalin and Ramakrishna." Back in New York, Hugh Guiler was spending his nonbanking hours learning to be an artist, making copper and wood engravings. These would serve as illustrations—he signed them "Ian Hugo," and the new name crops up now and then in the *Diary*—for his wife's writings, which she had begun to publish, since no one else would, on a hand-set press of her own.

The homemade books attracted little notice until a review of *Under a Glass Bell,* a slim collection of Nin's stories, appeared in *The New Yorker* in April 1944, above the august name of Edmund Wilson. "The unpublished diary of Anaïs Nin has long been a legend of the literary world," Wilson began. While her published work had been "a little disappointing," he noted, and her current book suffered from "an hallucinatory vein of writing which the Surrealists have overdone," it did yet show her to be "a very good artist." Wilson singled out a few stories to praise—one, "Birth," was about a miscarriage—and described the whole collection as taking place in "a special world, a world of feminine perception and fancy." He concluded with the information that the book was available at the Gotham Book Mart, and listed the address: "It is well worth the trouble of sending for."

Wilson later told Nin that he had noticed her years before at, in fact, the Gotham Book Mart, and had thought her, she records, "the most exquisite woman I had ever seen." When he and Mary McCarthy quarrelled, McCarthy had accused him of ("among many other things," Wilson said) being in love with Nin. In 1944, separated from McCarthy, Wilson began a kind of courtship by review, and he was far more successful in selling Nin's books than Miller had been. (Miller, stung again to jealousy, wrote from California, "So that did the trick! It's all so bloody spurious, what makes success here." And, in high Miller form, he reminded her of his contribution to her glory: "Rereading my own words about you I was so stirred that I was beside myself.")

Wilson and Nin began to see each other often, and her portrayal of this supremely lucid intelligence torn by loneliness and buffaloed by desire is genuinely touching, presented as it is without any sentimentality, or even any sympathy for Wilson, since Nin disliked him and thought his appreciation only her due. In Nin's diary, he pursues her while crying out "Don't desert me. Don't leave me alone," and she is forever fleeing, although not too far. She knows what he has done and can still do, "either by a destructive review or by indifference." He tells her that "I would love to be married to you, and I would teach you to write"—a dreadful insult, made worse by its heartfelt sincerity. And in a gesture of exquisite futility, he sends her a set of Jane Austen, accompanied by flowers and a note. "He was hoping I would learn how to write a novel from reading her!" Nin's calm retort: "But I am not an imitator of past styles."

She must have been—does not every reader mouth the words at this point in the *Diary*?—simply astonishing in bed. To what else can reason appeal? It may take a few more of the unexpurgated volumes to get to a full disclosure of her methods at the time, but it is clear even from the old, certainly expurgated volume four—in which she never yields to the ardent Wilson—that there had been a change: the ideal of the overmastering older man had given way, as she reached forty, to the ideal of her own mastery of a worshipful younger one. "The sons, the boys, the young artists are beautiful, delicate, subtle, tender, imaginative," she tells the diary, and she lived now surrounded by a host of them, serving them breakfast and being more understanding than their mothers, and interrupted only occasionally by a detective employed by frantic parents whose son had left home or school—these really were boys, some under eighteen—to learn wisdom at her hem. (In the fiction she was writing in the forties, all of it reworkings of her diary, a detective breaks in with the line "Having breakfast, eh!?" After he departs, the scene, in *Children of the Albatross*, continues: "Where sadness had carved rich caverns he sank his youthful thrusts grasping endless sources of warmth.") Nin especially resented Wilson's disparagement of her pleasure in these boys' company, "as if they were not the proper companions for a mature woman."

Wilson continued to cover Nin's work in *The New Yorker* for two years—until 1946—in reviews that are studies in constrained ambivalence. A book called *This Hunger* has "not much craftsmanship" and displays a "solemn, hieratic tone" that the author seems to have caught from Lawrence, but is nevertheless important "because it explores a new realm of material": that is, "the feminine point of view." Femininity appears once again as a virtue of last resort, a point of praise where there is really nothing to praise and yet something must be found. Wilson closed by suggesting that a commercial publisher should bring Nin's work to "a general public."

She got her publisher, and a lot of critical attention besides. For a while, it appeared that parts of the diary might actually make it into commercial print, thanks to a smitten young editor at Dutton named Gore Vidal. She had some success speaking on college campuses, and a handful of reviewers—mostly male—fell in with the notion of her representational femininity. Several reviewers singled out the "Birth" story as a particular example of her gifts or an exception to her faults. But the larger response to this first real scrutiny by impersonal arbiters was scathing. Nin copied out in her diary "the ugliest, most prejudiced" review, by Elizabeth Hardwick, which began with the observation that "no writer I can think of has more passionately embraced thin air" and concluded with the terms "vague, dreamy, mercilessly pretentious," and "a great bore."

Nin soon lost her commercial publisher, and the number of reviews of new works dwindled. The late forties and the fifties were for her a time of corroding anger and frustration, of constant but profitless attempts "to keep my work from being buried alive." She lists eleven publishers who turned down her novel *A Spy in the House of Love;* Putnam's called it "almost pornographic," and Macmillan, she adds, "said it was *esoteric.*" After she gave a public reading of her work, W. H. Auden was heard to ask, "What's with Anaïs Nin?" to which a friend of hers replied, "She is a poet." Nin writes, "Auden should have known this."

The *Diary* entries of the fifties are comparatively sparse, filled out with travel, with sunsets in Acapulco or California. The reason for the lack of published detail can be found in letters Nin wrote in 1950 to Kathryn Winslow, who ran a Chicago bookshop and gallery devoted entirely to the work—books and watercolors—of Henry Miller. Nin, as Winslow tells us in *Full of Life,* her memoir of Miller, was trying to sell her own manuscripts ("As you will see, *Winter of Artifice* was edited by Henry Miller") for fifty dollars apiece, or five hundred dollars for the lot. She claimed an urgent need of funds. Winslow obligingly mounted an exhibition, and Nin, whose husband had now taken up filmmaking (the films were short, and abstract, and "in the mood of my work," Nin assured Winslow), offered to come to Chicago for a reading and a film showing to help the sale. But she was travelling so much that she had to cancel, and finally wrote, "The truth is, dear Kathryn, I am caught in a dual life and my 'duties' at each of my two lives two loves are so tremendous that I can barely finish what I must do here for Hugo before I must rush back to take care of my life in California." A few weeks later she added, "As you know one half of my life is always the one I have to support, and only the half here with Hugo is the one I do not have to work for—so I'll be glad of the five hundred." After almost twenty years, nothing in the way she managed her affairs had changed.

Only the country was changing. In 1959 Grove Press announced the publication of the thirty-one-year-old *Lady Chatterley's Lover,* still barred from the United States mails. It was a national invitation to a trial for obscenity and to a redefinition of the term. The young owner of Grove, Barney Rosset, had written a college paper at Swarthmore called "Henry Miller Versus 'Our Way of Life,' " and since 1958 he had been trying to persuade Miller to let him publish *Tropic of Cancer.* But Miller, whose income derived almost entirely from sales of his books abroad, thought his own country not yet ready: "One has to die first, if you notice, before the ball gets rolling." He was living a quiet, isolated life on the Northern California coast: failing again in marriage but devoted to his children, steeped in Eastern mysticism, accustomed to his marginal status. He was no longer a man spoiling for a fight. And so, more than ten years after the first Kinsey Report shook America by putting into print what people had always been doing, American critics took the stand to defend D. H. Lawrence as an artist and as a man of religious vision whose book was a celebration of marriage. The judges were persuaded, and, cleared of her bad name, *Lady Chatterley* sold more than six million copies by 1960, mostly to people who hoped the judges had been wrong.

The next year, Miller decided to accept Rosset's offer. The story of the trial that ensued—or, rather, the trials, for there were at one point more than sixty cases pending throughout the United States—has been well told by E. R. Hutchison in *Tropic of Cancer on Trial,* put out by the victorious Grove Press in the late sixties. Where the Lawrence trial proved that *Lady Chatterley* was not obscene, the Miller trials proved, to all serious intents and purposes, that nothing was. Neither "degraded" subject matter nor offensive language could bar a work of art from First Amendment protection; Charles Rembar, the lawyer who masterminded both the *Lady Chatterley* and the *Tropic of Cancer* defenses, called his book on the subject *The End of Obscenity.* Although many stores and big chains, including Doubleday, Brentano's, and Scribner's, refused to carry *Tropic of Cancer,* it was a best-seller in 1961—No. 6 for the year, right behind *The Carpetbaggers*—and it was joined before long by the far raunchier *Tropic of Capricorn.*

For almost two decades, admirers of the freedoms in Miller's books and in his way of living had made the pilgrimage to his spartan cabin at Big Sur: from Frieda Lawrence in the forties—a period when *Harper's Bazaar* trumpeted "The New Cult of Sex and Anarchy" in describing his refuge there—to the Beats in the fifties (Kerouac got drunk on the way and never arrived) to students and tourists in the early sixties, when the place became so overrun that he moved out and with his big new money bought a big new house in Pacific Palisades. A cycle had been completed. The sexual liberty, the dabblings in Eastern religion, the nonconformism and anti-materialism that Miller had quietly unpacked from his battered thirties baggage along with his astrology charts and his *I Ching* were now the property of a whole segment of the population. Miller became the hero of a culture that he had in some measure created (and didn't entirely approve of), and he became, rather suddenly, one of the best-known writers in America. In the summer of 1962, Anaïs Nin arrived at his new front door, after fifteen years, with a friendly proposal to make.

The landmark 1961 edition of *Tropic of Cancer* had included Nin's preface and, in a new introduction by Karl Shapiro, a reference to "the journal of Anaïs Nin which has never been published," but which "Miller and other writers" swore to be "one of the masterpieces of the twentieth century." Miller was nothing if not loyal to his friends. In the uproar over his book these words had again aroused curiosity, but publication was still some distance away, and Nin now proposed to pave it with a selection of the letters he had sent her years before. Miller ceded all rights to her, and in 1965, the year that the two *Tropics* and the trio he called *The Rosy Crucifixion*—*Sexus, Plexus,* and *Nexus*—held five spots on *Publishers Weekly*'s lists of best-sellers, Putnam's brought out *Henry Miller Letters to Anaïs Nin,* which, as she records, "caused a stir and brought reviews and letters." (*The New Yorker* was stirred to pronounce it "one of the silliest books of the year." *Time* wasn't stirred to anything at all, although not necessarily because of its "practice to assassinate Henry and to ignore me.") Finally, in the spring of 1966, the Swallow Press and Harcourt, Brace & World jointly published volume one of *The Diary of Anaïs Nin,* beginning with the *annus Milleris* 1931 and representing, according to the introduction, about half of the material of her manuscript volumes thirty through forty, out of a total of a hundred and fifty volumes, or fifteen thousand pages of typewritten transcript—or, as would be specified later, in terms commingling Proustian largesse and Mozartian perfection, "thirty-five thousand hand-written pages with no erasures and almost no corrections."

She was sixty-three years old and "like a new woman, born with the publication of the diary." The old "neurotic plant," nipped and arranged just so, was looking like a flower garden. She was vindicated, she was famous, and her days—there was a little less time for recording them now—were filled with fan mail and television appearances and lectures. While a share of notices did make mention of "unadulterated narcissism" or "monumental gush," the book was, in large part, taken for what it announced itself to be: "a modern woman's journey of self-discovery." Its author was acclaimed in precisely the "feminine" terms she had set. Miller himself obligingly restated in print his original notion that "no woman has ever written in like manner" (which would seem rather to negate the archetypal point, but then one must also wonder what readers of Nin's purified and self-sacrificing account made of his further observation, "If she has no moral scruples, it is because she has arrived at a state of grace").

In any case, with the publication of volume two in 1967, Nin was able to boast that people were no longer seeing her as a mere extension of Henry Miller but were interested in her for herself. By the time volume four appeared, she had become one of the most popular speakers on American campuses—in 1971 Nin gave commencement addresses at Reed and Bennington—and in a turn of cultural events nearly as spectacular as the publication of *Tropic of Cancer* ten years before, she found herself called upon quite often by the devoted young women who made up her audience to denounce and repudiate the most vehemently reviled of oppressive male authors of the day: Henry Miller.

"Women have not liked my books as you thought they would," Henry Miller said to Anaïs Nin back in 1936. "You were wrong in that." She wrote his words down, of course, along with her own comments: "It is true. Women do not like to be depoeticized, naturalized, treated unromantically, as purely sexual objects. I thought they would, that they were tired of idealization. I thought I was." Her new lover had shown her otherwise. A year later, reading pages of what was to be *Tropic of Capricorn,* she wrote Miller angrily, "Instead of investing each woman with a different face, you take pleasure in reducing all women to an aperture, to a biological sameness. That is not very interesting, I say, nor very much of an addition. It's a disease." Nin's letter to Miller was published in 1987, in a collection of their correspondence entitled *A Literate Passion;* but the protest had already been publicly mounted, to great effect, in Kate Millett's 1970 bombshell, *Sexual Politics.*

Among the writers whose work Millett analyzed for its misogyny, Miller took pride of place. Lawrence and Norman Mailer might be the grandfather and the son, but Miller was the reigning patriarch. The opening exhibit of *Sexual Politics* is a quotation from *Sexus,* written in the early forties, in which the narrator pulls the imperious Ida Verlaine, in her silk robe and stockings, down into the bathtub with him, where she behaves "just like a bitch in heat, biting me all over, panting, gasping, wriggling like a worm on the hook." Millet condemned the scene in terms of its "underlying issue of power," an issue which became more evident when she passed on to another episode, from *Black Spring,* in which the same Ida Verlaine is beaten with a razor strop by her long-cuckolded husband; although the scene is anomalous in Miller's work, to Millett it seemed representative. In a chapter devoted to Miller's output, Millett concluded that, although he was "surely one of the major figures of American literature living today," Miller was guilty of "virulent sexism" and represented not sexual freedom but the puritanical guilt and disgust—with sex, and therefore with women—of his American origins. Miller's ideal woman was a whore, and the only biographical exception to his rule that "women are things," Millett wrote, was his respect for the work of Anaïs Nin.

It is true that women in Miller's novels are often ridiculous, as lost and gone in their sexual feelings, their endless trills of orgasm, as men are more conventionally shown to be. If anything, they have too good a time, and they have it too easily: our tireless hero need do little more than show up and unbutton. ("I was always a bad suitor. . . . I wanted *her* to make the advances. No danger of her becoming too bold!") Miller's women are big pink cartoons in a cartoon world. (Except when they are sublime and frightening and the doors to creation: when "cunt" becomes "womb." See Norman Mailer for the relationship between male awe and hostility.) They are as shameless in demanding their pleasures as his men, and they are not made to suffer for it—although the case of Ida Verlaine does form a peculiar, ugly exception. But then Ida is punished not for sex but for sexual treachery. And she is, moreover, a clear portrait of the woman who, for Miller, exemplified the crime: Anaïs Nin, of course.

Even the name is perfect. And Ida Verlaine was, Miller writes, "just exactly the way her name sounded—pretty, vain, theatrical, faithless, spoiled, pampered, petted. Beautiful as a Dresden doll, only she had raven tresses and a Javanese slant to her soul. If she had a soul at all! Lived entirely in the body, in her senses, her desires—and she directed the show, the body show, with her tyrannical little will which poor Woodruff [read Hugh Guiler] translated as some monumental force of character." Miller's sense of guilt over Hugh Guiler, and his anger at what Nin recorded as "the amorality of women, of all women, of women like myself" didn't prevent him from sleeping in Guiler's bed, or from taking his money. His moral qualms merely gave him dreams of "H. giving A. a harmless little crack"—which he "rather enjoyed"—and led to the more elaborate fantasy of Nin's comeuppance in *Black Spring*, the only book he dedicated to her.

But in his work Miller *was* a moralist, even if by negation rather than by example—deliberately raw and indignant, in the tradition of Swift or of his own hero, Luis Buñuel. Millett wasn't wrong in seeing him as a disgruntled puritan; the grounds of her attack are not far different from the grounds of Norman Mailer's defense of Miller in *The Prisoner of Sex* and *Genius and Lust*. Miller's two most recent biographers also agree, and have gone even further in repairing his image. Mary V. Dearborn's *The Happiest Man Alive* and Robert Ferguson's *Henry Miller: A Life*, both intelligent and highly individual books, have been used as sources for some of the incidents described herein. While the authors differ widely in emphasis and detail, they arrive at a nearly identical conclusion: that Henry Miller was, under it all, "a true romantic," in Dearborn's words, and possessed of (was it only twenty-odd years since Millett?) "a chivalric nature."

In her appearances on the college circuit—she gave fifty-six lectures between September 1972 and May 1973—Nin tried not to speak directly about Miller, and deflected questions with the general statement that she didn't believe in "waging war on men." Women should concentrate instead on their own achievements, she counselled, and on learning "to seduce, to attract their men into working for their liberation." Standing at the lectern in a flowing dress, her face a powdered mask, her voice a whisper, Nin embodied the notion, either archaic or revolutionary, depending on how old you were and where you'd been living, that femininity was not a matter of disposable social conditioning but—as she'd written in support of Lawrence forty years before—a valuable and unchangeable essence. And it was an essence clearly compatible, as she presented it, with a free and adventurous life, made up of famous friends and beautiful clothes and love affairs, and of the uncompromising work of an artist. Volume one of the *Diary* records that the ravishing but loveless Countess Lucie, "who appears all the time in the pages of *Vogue*," had sought out our heroine to ask the secret by which "you have conquered life." Nin's reply was the same one she gave to her audience: "I live out my dreams."

Any woman who thought and wrote as Nin did could not be without detractors, even within a generation befogged by incense and marijuana and Kahlil Gibran. One of the chief joys of the Nin era was the parodies she inspired, like the Columbia Forum's 1974 *Diary of Bananas Ninny*, by Susan Previant Lee and Leonard Ross, which opens in 47 B.C. ("My togas are known throughout the Empire") and moves with due speed to Paris, 1793 ("The blade mirrors my constant restlessness. . . . Equality bores me, as do fraternity, monarchy, anarchy"), and onward. In 1970 Gore Vidal, whose infatuation with Nin was short-lived, wrote a novel, called *Two Sisters*, that features the breast-flashing Marietta Donegal, a writer "unique in all but talent" who favors the adjective "ensorcelled" and who is engaged on a century-spanning autobiography; Marietta's primary characteristic is "a hair-raising desire to be noticed at any cost."

Nin never forgave Vidal—St. Teresa would have found it hard to forgive Vidal—and her last published volume contains a copy of a letter to him in which she demands to know, "for recording in the Diary," why he had spattered her "with his venom." ("For recording in the Diary" became a phrase of vast new power for her in her later years, restoring the diary to the effective status of the long-ago seductions.) She also had her differences with those who took her message seriously enough to call it harmful—women she called "revolutionaries, the kind which guillotined anyone with clean nails"—and felt sufficiently strong in her position to give notice to such a "radical" contingent in another letter, also published in her final volume: "I will no longer accept criticism. And I feel it is not clever of the political women to alienate me . . . because I have a great influence on women who do not respond to *them*."

All she left out of her image was the prop that supported the whole facade: her husband and his income, which had bought the house and the clothes and a number of the men, and had made it possible for her to live the life of the independent *artiste*. Of all this she said nothing. In fact, for a woman famous for retailing her life in the world's most extensive diary, remarkably little was known about her. Vidal, reviewing her volume four in 1971, seems to have been merely confused about her marital history when he wrote that back in the forties she was living with a second rich husband. (Vidal relates his shock at having been told by Nin one day that he should allow her husband to pay for dinner, "because he was not a poor artist, as they had pretended, but a banker.") But Deirdre Bair's 1995 biography, *Anaïs Nin,* let us know just how nearly and dangerously if unknowingly truthful Vidal was.

It was actually not until 1955 that Nin acquired her second husband, an unemployed actor turned forest ranger named Rupert Pole, a walking Greek statue sixteen years her junior, with whom she'd been living, in California, for a substantial part of each year since the late 1940s. ("I am running away with a most beautiful man.") The central problem with this second marriage was that Nin had done nothing to extricate herself from her first one, emotionally or legally, and that her husband—that is to say, her first husband, Hugh Guiler—still more or less expected her to be living with him in New York. For Nin, the situation entailed a good deal of frantic travel back and forth (the flight then took twelve hours) and many letters; to each man she signed herself "Your Beat Generation Wife." Bair fills in all that Nin's cash-hungry letter to Kathryn Winslow implied: the 1955 wedding, held in the chambers of an Arizona justice, with the bride's eyes warily fixed on a copy of the Arizona Criminal Record; the years of shifting stories kept barely straight by means of a folder of file cards stored in a bicoastal handbag, ready for quick consultation under the guise of checking notes for her fiction; the annulment obtained by the Poles in 1966, just ahead of the IRS—it was the first year Nin ever made any money—after which everything went on just as before.

To all this Nin's biographer is marvellously sympathetic, seeing her subject as more sinned against than sinning, the victim of deliberately willed male obtusity and extraordinary wear and tear. "Only Anaïs suffered from the double deception," Bair writes. So why did she do it? "In short, Hugo gave her everything but the kind of sex she wanted, and that she got only from Rupert, who gave her a life she despised," Bair concludes, which seems a bit reductive even for Anaïs Nin. It is true that each man knew no more than he wanted to, and that when precious ignorance failed, at last, there was understanding and forgiveness; perhaps each man had as much of his wife as he required.

In 1977, the year Anaïs Nin died, of uterine cancer, at the age of seventy-three, Hugh Guiler (as Ian Hugo) made a nine-minute film called *Luminescence,* which his program notes describe as the story of "a fragmented woman" who "returns to the waters of her birth." Two years afterward, he completed another film, which featured a Balinese dancer, laser photography, and images of Anaïs Nin: he called it *Reborn.* But it was Rupert Pole—the junior Mellors Nin spent a quarter of a century fearing she would lose to a younger woman—who nursed her devotedly through her final, dreadful illness, and who was left in charge of what had become her considerable literary estate. In obituaries on either side of the country, the *New York Times* and the *Los Angeles Times* offered different names of the husband Nin had left bereaved, and both were right.

Nin did not live to see the publication of her first best-seller. She saw the galleys, though, and approved the dust jacket, on which a twenties photograph of a woman toying with her garter was printed on brown paper, to evoke the heady days of sex books smuggled in from Paris. *Delta of Venus* was made up of fifteen of the stories that she had written for a dollar a page, for an anonymous client, back in the forties; *Little Birds,* published two years later, in 1979, added thirteen more. Nin wrote in a preface to the earlier collection that she had been reluctant at first to have this material made public, out of concern that she had derived her style from male pornography—the only model generally available—and had therefore "compromised my feminine self." But on rereading the stories she recognized her intuitive use of "a woman's language" and "a woman's point of view," and changed her mind. These books made Nin famous all over again, and established her reputation as the fairy godmother of the new literary domain of women's erotica—the Frida Kahlo of sex.

One wonders whether Nin's publisher recognized the irony of affixing the standard disclaimer "This book is a work of fiction," with no "resemblance to actual events," to two volumes that so clearly repeat and rework the contents of the diary. Here are the "violently sensual" Elena, with a "face from another world," reading *Lady Chatterley's Lover* on the train and learning that "this was the nature of her hunger"; and Rango, "an American writer, whose work was so violent and sensual that it attracted women to him immediately." In the writing itself, there are in fact improvements. The client's instructions to "leave out the poetry" may have been the best editing advice Nin ever got; there is a minimum of the "Chance is the fool's name for fate" dialogue that marks the novels and the diary. (True, someone does tell Elena, "You draw me into the marvelous. Your smile keeps a mesmeric flow.") But what one notices, reading these stories from beginning to end, is that far from manifesting a new female freedom, or a "joyous display of the erotic imagination," as a cover quote promises, they are, in disconcerting proportion, tales of harrowing female frustration.

That Nin's notion of pornography should be school-of-Lawrence is hardly surprising—"she kneeled and prayed to this strange phallus which demanded only admiration," etc.—but there is nothing Lawrentian about the way the men in these stories, frequently "passive" or "wilted" or "frightened," or offended by a woman's forwardness, leave the women "unsatisfied," with "desire unanswered," or even choosing "the course of pretense." All the last-minute happily-ever-after "submissions" are no match for the pages of sexual suffering and humiliation these women undergo. Is the distinguishing characteristic of the "feminine" point of view a sense of sexual deprivation? In the part of the diary published as *Henry and June,* Nin reports that Miller startled her, early in their relationship, by telling her, "Your sensuality doesn't convince me." Dr. Allendy at one point accused her of being frigid, prompting admissions about failures with Henry and about only "occasionally" having an orgasm with Hugo; from Henry himself she conceals "the fact that I rarely get ultimate sexual satisfaction" despite her pleasures with him, and she writes in *Incest* that after "lovemaking" with her father "I did look transfigured, although I had felt nothing." Perhaps she worked it all out, and in any case, one doesn't wish to pry—one didn't wish ever to know *any* of this—but like the hidden husband, it does tend to call elements of the myth into question. In a story in *Little Birds* called "Mandra," a woman in analysis discovers "that she has never known a real orgasm, at thirty-four, after a sexual life that only an expert accountant could keep track of." As a solution, she tries "to awaken by falling into bed with anyone who invites her." Nin's comment, as narrator: "She deceives everybody, including herself."

In 1973 Nin blushingly reported that on a visit to her publisher, Harcourt Brace Jovanovich, the salesmen applauded when she entered the room. Small wonder. Today, what with the original *Diary,* the *Early Diary,* the erotica, and the new, unexpurgated volumes of the *Journal of Love* (drawn from material made available after Hugh Guiler's death, in 1985), Niniana has become a veritable industry. In an age in which women have so little time to record their own daily lives, Nin's publisher has provided a steady supply of what Oscar Wilde's Gwendolen—who, like Nin, never travelled without her diary—called "something sensational to read in the train."

Having passed through barrier after barrier, Anaïs Nin finally, with the publication of *Incest,* passed beyond even parody, although the reader cannot entirely reject the suspicion that this is Gore Vidal's brashest and funniest *coup de plume.* The volume includes twelve "ensorcelled"s or "ensorcellment"s—after three appear in the first pages, one can't help keeping track—and, more broadly, a multitude of passages such as the following, in which the author appears to be contemplating a globe: "When the earth turns, my legs open to the lava outpouring and my brain freezes in the arctic—or vice versa—but I must turn, and my legs will always open, even in the region of the midnight sun, for I do not wait for the night—I cannot wait for the night—I do not want to miss a single rhythm of its course, a single beat of its rhythm." Here and there, passages like this are marked by a note: "I place this immediately in my book."

It does not seem possible that the people who have ushered these lava outpourings into print have read the result. One may accept the writing at this point as a given, differing from Nin's other work in concentration rather than in kind, and concede that it is not a matter of historical importance whether she swallowed semen for the first time in 1932, as she lets us know in *Henry and June,* or in 1934, as she says here. But it certainly does matter to Otto Rank's reputation—if to nothing else—that the preface, by Rupert Pole of the Anaïs Nin Trust (and the publisher's publicity announcements derived therefrom), claims that it was under Rank's professional guidance that Nin set out to seduce her father, when by her own account, printed between the same covers, she began the incestuous affair in June 1933 and saw Rank for the first time the following November with the stated plan of receiving "absolution for my passion for my Father."

Incest and abortion, the big revelations of this volume, have been a matter of supposition for a long time. Mary V. Dearborn, listing men who might have been suspected of fathering Nin's stillborn child—Guiler, Miller, Rank, her cousin Eduardo Sanchez—includes "even, it was rumored, her own father," and adds that Nin herself encouraged the rumor. Robert Ferguson suggests that Nin "took steps to ensure that the child was born dead." Although Nin, who never had children, referred to her "stillbirth" as late as 1971, in 1972 she was one of fifty-three "respected women residents in the United States" to start off a *Ms.* petition demanding national abortion reform, which was printed in the magazine under the general heading "We Have Had Abortions."

Still, this abortion is shocking to read about. The story that Nin made of it, the celebrated "Birth," begins with the line " 'The child,' said the doctor, 'is dead,' " and goes on for its five brief pages to describe the agony of a woman stretched on a table, six months pregnant, too weak to push the child from her body and too tender of spirit to be fully willing to push it out, "even though it had died in me" and "even though it threatened my life." In the end, she bravely insists on seeing the baby—it is a girl, and the nurses want to hide it from her—and finds it, in her concluding line, "perfectly made, and all glistening with the waters of the womb."

The story was drawn from Nin's diary, and reappeared in elaborated but not substantially altered form as the climax of her volume one, published in 1966, not long after Sylvia Plath's almost equally excruciating birth scene in *The Bell Jar.* Among the elaborations were a doctor's warning that Nin would require a cesarean, and brooding paragraphs on the order of "There are too many men without hope and faith in the world. Too much work to do, too many to serve and care for. Already I have more than I can bear." There is the same well-evoked pain as in the story, the same noble insistence on seeing the child, the same wistful admiration—"Regrets, long dreams of what this little girl might have been"—and then resignation to the fact that "Nature connived to keep me a man's woman, and not a mother; not a mother to children but to men." For a coda Nin offered a religious epiphany, in which, the next morning, "God penetrated my whole body," reassuring her that "everything I had done was right." The entire "birth" episode gave a weight of seriousness to the first published volume of the *Diary,* and was partly accountable for that volume's being read in terms of female courage, and in terms of Nin's own rhetoric of the artist giving birth to herself: "I was born. I was born woman. . . . This joy which I found in the love of man, in creation, was completed by communion with God."

There were many clues in this account to what really happened, but they were easily ignored in the light of Nin's insistent claim to truth (which only underscored the commonsense presumption that, as Henry Miller put it, "to lie in a diary is the height of absurdity. One would have to be really insane to do that"). In the last volume that Nin published, she was still claiming, "I have not changed anything in the Diary, only omitted." *Incest,* then, gives us what was omitted, beginning with the author's absolute clarity of purpose from the moment of the discovery of her pregnancy, in May 1934: "I know it is Henry's child, not Hugh's, and I must destroy it." Instead of a doctor, there are treatments by the "sage-femme," or abortionist, but in early June she has "not yet ejected the unwanted child." The affair with Rank then takes over—"I feel that I am like a St. Theresa [*sic*] of love"—and the pregnancy is not mentioned again until mid-August, when she reports taking quinine "to rush the delivery of the Easter egg."

When this also fails to work, Nin finally sees a doctor, who finds the child "six months old, and alive and normal," and says that an operation will be necessary. The actual procedure is reported just as before, and culminates in the same brave outcry: "Show me the child!" But this later version includes more of what she sees: "It was like a doll, or an old miniature Indian. About one foot long. Skin on bones. No flesh. But completely formed," and yet also "almost nonhuman, not mental looking, a little bit monstrous." The convolution of lies and editing and reediting is hard to sort out, and here, still, are the luxuriantly sentimental phrases—"regrets, long dreams of what this little girl might have been," and "the simple human flowering denied to me because of the dream, again, the sacrifice to other forms of creation." This abortion was a sacrifice made to art, and to ensure "my destiny as the mistress, my life as a woman."

Sure enough, her bedside is soon thronged: with Hugh, Henry, Eduardo, and Rank, all of them "amazed by my appearance. The morning after the birth: pure complexion, luminous skin, shining eyes. Henry was overwhelmed. He was awed. . . . Eduardo brought me an orchid. The little nurse from the Midi left all her other patients waiting to comb my hair lovingly. All the nurses kissed and fondled me. I was bathing in love." Even God follows suit, in a demonstration worthy of that given, indeed, to St. Teresa, composed of light and ecstasy and penetration, and offering the joyous confirmation, here again, that "everything I had done was right."

This is a horrifying scene, if not in the way that Nin intended, and is perhaps, even now, a dangerous one. In a review of *Incest* published in the Sunday *Times* in 1993, Katha Pollitt wrote that the problem with Nin's abortion story was that she had not disclosed the truth years ago, at the time of the diary's original publication, when "it would have done some good for a well-known older woman to have gone public about her abortion." Is it imaginable that so brutal and frivolously self-serving an account could have contributed to the argument for abortion rights? The "stillbirth" of volume one appeared in 1966, a year before this country's first state legislatures voted to relax nineteenth-century restrictions on abortion. In this instance, rewriting her history was probably Nin's best deed for the feminist cause, and her most important lie. For even in an age of hard-won and vulnerable freedoms, the truth we are offered now is recognizably obscene.

While the fully revealed dramaturgy of the event tells us nothing essentially new about Anaïs Nin, it does refocus a career of trivial wickedness and familiar vice into something larger. Of course, there is no moral overview. Nin had made herself into a character rather than an author; that is perhaps what is implied when she is so often described as "self-invented"—a Madame Bovary sitting down with a pen and the notion "Flaubert, *c'est moi.*" But this is the most powerful, the most resounding scene she ever wrote, precisely because of the limits of her vision and the terrible palpability of a world where, as she put it in a moment of rare, grim honesty, "the only distinct personage is one's self." For the reader able to escape the solitary confinement of these endless pages through the mere act of closing a book—such a simple deliverance—relief is dulled only by a shuddering pity for the woman who lived all her days trapped inside.

The Strong Woman

Mae West

Because there were no available acting roles for a woman who drove men wild and enjoyed them in bed by the dozen and gave as good as she got and didn't want to marry and never suffered for any of it, Mae West had to become a writer before she could be a movie star. She began her literary career with a sketch for a vaudeville act in 1913, when she was twenty and her fame still rested largely on her ability to perform a well-advertised "muscle dance in a sitting position." By the time her first successful theatrical opus, entitled *Sex,* got her arrested in New York, in 1927, she'd been honing her playwriting skills alongside her nonpareil shimmy and cooch for over a decade. She was still, however, a work in progress: a low-down way to move in search of a philosophy, or maybe it was the other way around. By her movies of the thirties—several of which she wrote as well as starred in—style and content are so tightly joined that the comedy seems but a natural consequence of her remarkable bearing and physique: languorous yet corseted in armor, solid yet succulent, minutely activated with audible whirrings and purrings yet hugely, defiantly immobile. In an age of high kicks, high speeds, and easy smashups, Mae West was a woman structurally engineered to stand firmly on her own two feet.

A new wave of interest in West as a kind of precocious modernist has resulted in a number of books concerned with her formidable feminism, her games with what she would never have wished to call "gender," and her enduring status in the American pop pantheon. These largely academic studies differ in focus, but not in basic biographical content, from the old picture-book stories of West's career. All accounts agree that her breakthrough creation was the Gay Nineties saloon singer "Diamond Lil," a role that brought her to sudden and stunning respectability when her play of the same name opened on Broadway in 1928. Overnight, the woman who had been dismissed as too vulgar for big-time vaudeville and jailed for violating public morals was transformed into America's leading actress-playwright, "more admired by her public than is Jane Cowl, Lynn Fontanne, Helen Hayes or Eva Le Gallienne," according to the *Herald Tribune.* The play's Bowery-bar setting made for some thrilling intellectual slumming, and West's one-hip-at-a-time portrayal of an implacably good-natured quasi-prostitute overfond of her diamonds and her work was hailed as a theatrical archetype comparable to Harlequin or Pierrot, a portrait as grandly abstract and theatrical as Bernhardt's Tosca. We know that West had finally found her distinctive self—as distinctive as Chaplin's Tramp or Groucho's Groucho—because Mae West imitators were already finding work.

But still Hollywood was afraid to touch her. It was only after West had entirely rewritten her part and stirred up some life in an otherwise dead-on-arrival George Raft movie, *Night After Night,* in 1932—"Goodness, what beautiful diamonds," the hatcheck girl simpers; "Goodness had nothing to do with it, dearie," Mae instructs—that Paramount agreed to take a chance on *Diamond Lil.* Studio censors tore through the play, eliminating a little subplot about white slavery, a homosexual characterization, and such provocative lines as "the last guy she had." Their goal was not so much to eliminate sex as to erase all signs that the leading lady derived either pleasure or money from the experience. After months of haggling, the red-inked script was finally shot in eighteen days. The project was retitled *She Done Him Wrong* and the heroine's name changed to Lady Lou—to sidestep the notoriety of West's loot-bedecked Lil and to suggest her morally and socially elevated condition.

Censors were still so skittish that a week before the premiere Paramount recalled the already distributed prints in order to disembowel one of the "dirty blues" numbers that West had freely incorporated into Lou's 1890s act. (Cut to just the first and last verses, the deliciously insinuating "A Guy What Takes His Time" was no less clear in its meaning, only manically abrupt in its delivery—less a sexual rhapsody than a manifesto on the joys of foreplay.) Yet in spite of all the studio's efforts to clean up and tone down West's act, *She Done Him Wrong* was a huge box office hit: enormous crowds, late-night screenings, return engagements. Mae West was a top star of 1933, and a national fixation ("She's as hot an issue as Hitler," *Variety* wrote). And Diamond Lil—with her two-handled curves, her speakeasy wisdom, her armor-plated heart of gold—was a new American folk heroine, a part of our mythic history as surely as Betsy Ross. ("And all *she* ever made was a flag.")

It's hard for anyone watching *She Done Him Wrong* today to believe that West was ever really considered either dangerous or desirable. Many viewers point to the fact that she was almost forty when she came to the screen (if admitting to twenty-nine). But West had never traded on conventional beauty. An early notice, when she was just eighteen, refers to her simply as "one of the many freak persons of the vaudeville stage." (She appeared on a circuit that also proudly billed the Half Woman and Venus on Wheels.) In her theatrical roles, she veered between blunt sexual display—especially when she was trying to be serious, in which case reviewers consistently found her crude and fat—and a far more successful comic raunchiness based in burlesque. Sex was her subject, not her effect. And it was what she had to say about sex that was genuinely dangerous. As West's admirer and fellow aphorist George Bernard Shaw once observed, "If you tell people the truth, make them laugh or they'll kill you." It required all her accumulated veils of wit and rouge—and about thirty years of stage experience, starting at age seven—to pull it off.

Mae West, an ambitious woman, began life as an ambitious woman's daughter. Matilda (Tillie) Doelger came to the United States from Germany in the 1880s and married an Irish boxer named John West. Like good Americans, they settled in working-class Brooklyn, had four children, and became obsessed with show business. A firstborn girl died in infancy; the next, Mary Jane—born in 1893 and known almost from the start as Mae—strongly resented her father, who had wanted her to be a boy. In particular, she recalled that she didn't like his cigars and she didn't like him to touch her ("Freud wasn't there to explain it to me"). But Mae worshipped her mother, who worshipped vaudeville, and whose life was soon entirely devoted to making her talented favorite a star.

Mae was on the boards in local theatres a few years after she could walk, playing in *Little Lord Fauntleroy* and as Little Eva in *Uncle Tom's Cabin,* and performing as a "coon shouter" (probably in blackface) between the acts. Distractions like school—and, later, romance—were quickly left behind. Confronted with the girl's rising adolescent emotions, her mother warned "that I could use them to help me be very famous or I could waste them on the first man," West explained in an interview in 1934, by which time the first man was a very distant memory. All potential attachments were summarily ended. West would sometimes claim that she'd been broken of the ability to love; or else that she'd prudently chosen to love nobody as she loved herself. It seems true that the only great romance of her life was with her mother. (West did actually marry once, at seventeen, without Tillie's knowledge, as a cover for a possible pregnancy. The couple never even lived together, and she walked off forgetting all about it.) It was from Tillie that Mae learned the all-important lesson that "one man was about the same as another." And so, "I learned to take 'em for what they were. Stepping stones." It was to this absolute, unswerving focus that West attributed her success.

Her first big moneymaker came in 1926, when the three-act *Sex* opened at Daly's Sixty-third Street Theatre. In search of theatrical legitimacy, she wrote the play under a pseudonym. Then she rented the theatre, hired the director, and played the leading role: Margy LaMont, a bitter working-class prostitute out to better her lot. No diamonds, no plumes, no multiple entendres. West's awkward script was full of determinedly hard-hitting social truths, and its tone was blatantly angry. "Why, ever since I've been old enough to know about sex, I've looked at men as hunters," Margy announces. "All the bad that's in me has been put there by men." She further explains, "I began to hate every one of them, hated them, used them for what I could get out of them, and then laughed at them." In her later years, West hoped that the play would be revived as a masterpiece of sexual honesty before its time, like Ibsen's *Ghosts.*

The review in the *Mirror* bore the headline MONSTROSITY PLUCKED FROM GARBAGE CAN, DESTINED TO SEWER. The reviewer moaned that his task should have gone to the Health Department, while another wrote that the audience had been left with "that 'dark brown' taste which results from proximity to anything indescribably filthy." But, thanks to West's raw illustrations of her clearly stated subject—"lust—stark, naked lust," clucked the *Herald Tribune*—*Sex* sold out for months.

It was while she was having her clothes profitably torn off nightly that West approached an even more sensational subject, in a play she called *The Drag.* Although it was the first show she'd dreamed up without a role for herself, this "homosexual comedy-drama" came closer to West's recognizable, full-blown image than anything she had done so far. She later claimed that she'd been drawn to the subject when her pursuit of an apparent "he-man" met with a less than fully heterosexual response; this astounding experience set her—she said—to reading Freud. In fact, the plot of *The Drag* was pure, hard-boiled sociology culminating in a plea for tolerance—that is, for regarding homosexuality not as a crime but as an illness. West's dramatic finale, however, was a thirty-minute drag ball, largely improvised, in which giddily self-described "queens" and "queers" in gowns and floating boas performed—in a manner even then called "camping"—passable impressions of a Mae West who had not yet quite come into being.

The legal reaction was prompt. West was arrested the night after the new show's unadvertised New York preview. She was prosecuted, however, for her production of *Sex*—still running strong—and eventually spent eight days in the lockup on Welfare Island, well photographed as she arrived and left, and put out mostly by the regulation cotton underwear. Within months, the State Legislature had outlawed any depiction of homosexuality on the New York stage. Yet what West disastrously if inadvertently cost the gay community in terms of open theatrical representation, she repaid during the rest of her career in representations of Mae West. It was from *The Drag* and from its spin-off, entitled *The Pleasure Man* (raided and closed in October 1928), that she added the final, transforming touches of exaggeration and irony—that is, of conscious "camping"—to her mythic self.

The most reliably sober and informative of the recent crop of West studies, Marybeth Hamilton's *When I'm Bad, I'm Better: Mae West, Sex, and American Entertainment,* pinpoints West's startling transformation from frank man-hating to easy sexual mockery as occurring during rehearsals of *The Drag* and *The Pleasure Man,* when West was surrounded by a crew of improvising drag queens while revising the script of *Diamond Lil.* Before the rehearsals, Lil on the page was hostile, angry, badly used. And afterward: there stands the bulletproof, shimmering, vastly amused being whom the critic Parker Tyler once characterized as the reconciliation in a single body of the gay son and his gaily painted, warmly all-forgiving mother.

▪ ▪ ▪

One pauses before the delicate question of where the anger went. *The Pleasure Man,* despite bouts of hysterical buoyancy, is a bizarrely vengeful melodrama, in which the cruel Don Juan–like hero of the title is castrated by the brother of one of the women he has ruined, and dies of his wounds. ("If you're a man, thank God I'm a female impersonator," one character tells him.) In a far happier fashion, *She Done Him Wrong* amounts to a bright Cinderella version of what remains nevertheless a dream of revenge. "Men's all alike, married or single. It's *their* game. I happen to be smart enough to play it their way. You'll come to it," Lady Lou reassures a young woman who has tried to kill herself after being deserted by a married man. There's a reason that for her second big picture—*I'm No Angel* (1933)—West's audiences were made up largely of young women. She was teaching the same invaluable lessons she'd learned at her mother's knee: get the smug little prince to wait his turn in line and make sure you can pay for the dress and the slippers yourself.

Fittingly, it was from her mother, too, that West finally found the way to give her ambitions shape and meaning. Tillie had died in 1930, while West was playing in *Diamond Lil;* several shows had to be cancelled, as the star could not speak for days. (For the rest of her life, West kept with her a photograph of her mother that visitors were surprised to note had been crudely retouched with layers of West's own painstakingly applied cosmetics.) In 1959 West dedicated her autobiography *Goodness Had Nothing to Do with It,* "In loving memory of my MOTHER without whom I might have been somebody else." She wrote of Tillie as a great beauty, "sexy, but refined," and so shapely that she'd worked as a corset model. But Tillie's allure was a matter of far more than looks. "There was a power and vitality about Mother that made a man melt before her glance." Her husband had called her Champagne Til, and as such she'd been the inspiration for West's omnipotently attractive alter ego, who'd started out as Diamond Til.

Each of the vital components of West's new persona has been well documented. The traditional biographers have always pointed to West's own statements about her mother; the authors of "cultural studies" have examined the elements of drag. But surely the two are intimately connected. The drag-queen touch that went into the creation of Lil was not simply a matter of an appealingly outrageous or novel style. West had been familiar with that style since her early vaudeville days, and she'd known all the pros—including Bert Savoy, lauded by Edmund Wilson as "a gigantic red-haired harlot," whose tag line "You must come over" seems to have stayed with her. Only in 1927 did she finally find the style really useful, when it appeared that the qualities that men brought to playing women gave her what she needed for her awed, adoring characterization of Til: detachment, control, coolness, laughter. One might say: impregnability. The exaggerated profusion of laces and plumes and jewels and the towering picture hats—the period styles of the real Til's own glamorous prime—make a sublimely crazed rendering of femininity that suggests not only a man pretending to be a woman but a little girl pretending to be the most gorgeous grown-up woman she'd ever seen. Small surprise that the only other female in films who displayed a sense of power as unapologetic and resolute as West's was Shirley Temple, and that the two were paired as box office favorites throughout their brief, wonderlandish reign.

Five of the eight movies West made between 1933 and 1940—the essential span of her career—were set in the Gay Nineties, not otherwise a terribly popular period. This was more than the usual Hollywood custom of repeating a successful formula. By contemporary standards, West was a hefty woman, and it was the period dress that gave her an aura of beauty: carefully draped and shaped by whalebone, she was suddenly curvaceous instead of fat, provocatively contained instead of unintentionally overexposed (which is how she looks in the photographs of *Sex* and her other early shows). Even her non-period films managed to keep her poured into a more or less nineties silhouette. Basically, after 1928 no one ever saw her legs again.

No sooner had she hit the screen than her extravagantly sexual redefinition of what had become a resignedly maternal shape spurred an insurrection among "ordinary" women against postflapper fashions, so famously free yet also demanding the straight lines of a near adolescent female body. In September 1933 *Vogue* billed a story on West and the corset as "a return to normal" and "a lady's way of saying that the depression is over." Flesh was back. In Paris, "a little pinching . . . and bulging" à la Mae West—that thrilling savage, possibly a blond Negro—was the rage. Colette herself protested vigorously upon the release of *I'm No Angel* that it was "an unforgivable detail, a violation of principle" for Mae West to have lost a little weight.

In her spectacular good health and amplitude and appetite, West seemed to beckon her audience away from all that was modern and angular and neurotic—at times, very nearly away from the decline of civilization itself. "She may not be an answer to Spengler," a reporter for the *World-Telegram* opined in 1933, but she would do for the moment. Mussolini advertised her as a kind of fertility idol (despite her known childlessness) to encourage Italians to have large families. In Vienna in 1934, her films were banned after church groups and conservative newspapers protested the fact that citizens of newly de-Socialized "Christian Vienna" were waiting in long lines to see such displays of immorality. That same year, the gods of Hollywood, also horrified by the immorality of the Mae West pictures that had pulled Paramount back from the edge of bankruptcy and made her the biggest star in the country, agreed to strengthen the infamous Production Code. The code cut sexual references to nil and required that in the end the audience feel that "evil is wrong and good is right." With the slow, sure stranglehold of puritan hypocrisy, it made West safe for an America that had never wanted her that way. "You can't be funny if you have to be clean," she'd sworn as early as 1933. And she couldn't be—not with the things she had to say. At the very moment the studio was publicizing her formidable power as a "one-woman production staff"—she wrote scripts and monitored direction and costumes and music—the censors were stripping her of any power to make it all matter.

Belle of the Nineties, released in 1934, was based on *The Constant Sinner,* a novel she'd written—dictated, actually—in 1930. Reissued in the mid-1990s by Virago Press (along with West's novelization of *She Done Him Wrong*) the book is still surprisingly fresh and harsh. It tells the story of a ruthless whore who climbs her way up through the New York underworld, not in the Gay Nineties but in the rough 1920s, mostly in morphine-and-cocaine-suffused Harlem nightspots and in the ringside boxing world. West's heroine, Babe Gordon, is something of a pugilist herself: "Every man she looked at she sized up as a fighter would an opponent. How would she handle him, out-smart him, out-point him?" Babe, who enjoys her work, selects the men she enjoys most from the boxing ring—"She could get an idea, from the way a fighter handled himself, in the ring, of the way he treated women"—and one of the book's major themes is the sexual attraction between white women and black men, a matter of which the author, by all accounts, spoke from personal knowledge.

> "How in the name of all that's decent, Jack, could a woman like that . . . allow a black to make love to her?"
>
> ". . . *Sexual preference,*" replied Jack.

The novel's denouement is a wicked parody of the famous racial rape in *The Birth of a Nation.* After the golden heroine has enjoyed a long afternoon in bed with her black gangster lover, another of her lovers bursts into the room and shoots him dead; her husband, who agrees to take the rap, is acquitted for having defended his innocent wife against the violent defilements of "the black gorilla" and for having "upheld the best traditions of the white race, the honour of its womanhood."

The book sold well—more than ninety thousand copies in two months—and West turned it into a play, which hit Broadway in 1931. Ironically, and against her apparent wishes, the gangster was played by a white actor in blackface. At every curtain call, he removed his wig with a flourish—a gesture borrowed from female impersonators concluding a vaudeville act—to expose a band of bare white at the hairline and make it clear that no actual black flesh had ever come in contact with white.

Needless to say, there was nothing left of West's original intentions by the time the movie was made. Period, plot, characters, and of course the race factor disappeared entirely. West played not a modern whore but an 1890s saloon singer, and the only notable black speaking role went to the maid. Even so, West did win one battle: she got Paramount to hire Duke Ellington and his musicians instead of a studio band. (It was Zora Neale Hurston who noted, after seeing *Sex* in 1926, that West's musical tastes—she'd played "Honey Let Yo' Drawers Hang Low" on the piano—were strongly, traditionally black.) And, via the cinematography of Karl Strüss, there is one number that seems to evoke something of the forbidden subject she had written about. As West stands on a balcony singing the bluesy "Troubled Waters"—"Oh, I'm gonna drown in those troubled waters, / They're creepin' round my soul"—a close-up of her bejeweled platinum-blond head as she warbles out her pain and longing is superimposed on a view of a black revival meeting at a campfire, the figures leaping and shouting ecstatically around the flames. This particular scene was cut when the film was shown in England. Enough of the rest of the film was cut before its release in the United States to make the confounding muddle of a story the focus of reviews.

The censorship battle went down to its dirtiest in 1936, when William Randolph Hearst set his empire of newspapers on *Klondike Annie*—the mildest of West's brews, in which her usual irreverent character reforms and becomes a sincere, if somewhat unorthodox, missionary. (The studio had been especially pleased with the result, after scissoring out lines like "You can't save a man's soul if you don't get close to him. It's the personal touch that counts.") But the subject of religion was a lit match. Under such headlines as THE SCREEN MUST NOT RETURN TO LEWDNESS, Mae West was declared uniquely responsible for "the uprising of the churches and the moral elements of the community against the filth in motion pictures." Couldn't Congress do something about her? She was "a menace to the Sacred Institution of the American Family."

She was also at that time the highest-paid woman in America, at $480,833 a year, and the second-highest-salaried American of either sex, the first being William Randolph Hearst. It isn't hard to speculate that Hearst's role as owner of a rival studio, Cosmopolitan Productions, drove him to his campaign. But West had her own ideas: "No wonder Mr. Hearst and his high clean living moral values was writing editorials against me," she wrote in her autobiography. "He hated to see a woman in his class."

Paramount's self-defense was complicated by West's tenacious battle against every change in every film: she'd actually smuggled out to theatres a print of *Klondike Annie* that had escaped the studio's final cuts. Still more problematic was the fact that even after a script had been torn apart she could make the barest participle left on a page sound oozily lubricious. All she required was a little bit of unscriptable pressure laid against a word, or a swivel down somewhere in the locomotive axis, or, at most, the famous deep, cooed "Hmmmm?" Torpedoed content bubbled right back up in innuendo and undulation. ("A little momentum to remember me by.")

Hopelessly beleaguered, the studio dropped her contract in 1936. But the situation was no easier with the independently produced *Every Day's a Holiday,* made in 1938. With Paramount still controlling distribution, she fought the good fight over lines as timid as "I wouldn't even lift my veil for that guy." This was her first movie to lose money. With so little actual material left, not even her near Kabuki stylizations could get her point across anymore. She was now official box office poison. In 1939 she made the last major film of her career, the overstuffed and overpraised *My Little Chickadee,* in which she played a living effigy of herself opposite W. C. Fields: the two old troupers exchanged dead references and stale quotations where warmth and wit had thrived.

She remained world-famous, but as an artifact, a cartoon, an animated shape: in 1940 the RAF named its bulbously inflatable life jacket a "Mae West." Watched over by a manager handpicked by her mother, she kept to herself in her white-and-gold Hollywood apartment—Louis XVI by way of an old movie set—closely attended by a newfound spiritual advisor. She had never taken part in the Hollywood social scene, and her circle remained a shifting group of fighters and gamblers and old New York cronies. She'd moved her father out to be with her in California—eventually to the same apartment building—and she took care of an alcoholic younger sister, who'd spent her life not being Mae West. Ever the star, she continued to be photographed on her canopied bed under her famous mirrored ceiling, and what one might call her romantic life went on just as she'd always advertised: a fantasy in which the subject was sex and the multiplied reflections were dazzling.

For those who find it hard to believe that a woman could ever really pursue such an existence, several long-term quasi-marital relationships have been put forward: with her manager, James Timony, who handled her business from Tillie's time until his death in 1954; with the former muscleman Paul Novak, who picked up where Timony left off and stayed on for the rest of her life; and, discreetly, with her chauffeur, the onetime boxer Chalky Wright. But while these relationships may have provided some degree of emotional stability, they were not love affairs—or, if they began that way, they didn't remain so for long. Nothing ever did. Whether she herself pronounced this to be a triumph or—occasionally—a tragedy depended on her mood, her audience, and the year.

She made one more film attempt, in 1943—a turkey called *The Heat's On*—and then appeared on Broadway, briefly, in a play she wrote about Catherine the Great, in which she staggered her audience perhaps more than ever by playing the rapacious empress entirely straight. (She said she'd got all the experience she needed at Paramount.) In 1948, when she was fifty-five, Billy Wilder virtually begged her to play the lead in a new movie that he and Charles Brackett were writing, which would be called *Sunset Boulevard*. She refused even to look at an outline. She was incensed at being asked to play a has-been or an older woman: she was at her peak; she could pass for twenty-six. She chose instead to spend four years on the road in another production of *Diamond Lil*.

"She alone, out of an enormous and dull catalogue of heroines, does not get married at the end of the film, does not die, does not take the road to exile, does not gaze sadly at her declining youth in a silver framed mirror in the worst possible taste," Colette wrote of Mae West in 1938, "and she alone does not experience the bitterness of the abandoned 'older woman.' " Colette, then sixty-five and a Commandeur de la Légion d'Honneur, had been a "mime artist" in low-rent Paris revues at about the time West was growing out of Little Eva. Into her fifties, Colette had appeared onstage as her own most moving heroine—the courtesan Léa, of *Chéri,* who loses her young lover because of the agonizing encroachments of age upon her beauty. No writer was more expert in the sacrifices a woman makes and rues for *l'amour.*

Colette was also a literary specialist in the long tradition of actresses and singers and dancers who were able to go it alone—since she believed that only in these métiers had women found independence, accomplishment, and adoration enough to replace the humbling needs of women more obscurely employed. (Certainly, history and literature tend to support this view. So early and well established was the notion of the stage diva's freedom from the standard lot of womankind that George Eliot, writing in 1876, portrayed the opera singer "Alcharisi," in *Daniel Deronda,* as imperiously disdaining marriage and children until her voice begins to fail. Simone de Beauvoir—like Colette, a West devotée—exempts this highly specialized class of women from the amatory afflictions of the second sex.) One of Colette's idols was Sarah Bernhardt, who had simulated so well "the feminine suffering to which the theater had rendered her immune." Another was Mae West, who simulated nothing.

But if West really was neither sad nor bitter as she grew older, the alternatives she presents are not reassuring. What happens to a highly sexual woman past a certain age—at Colette's last view of West, she was a venerable forty-five—who rejects the "sexless dignity" that both de Beauvoir and Colette prescribed as the only way to live on contentedly? (Prescribed for others, it must be said, since both of them subsequently fell into long relationships with much younger men.) Where in this feminist age are the unashamed counterexamples? Harlow died at twenty-six and Monroe at thirty-six; Garbo went into hiding before she was forty; Goldie Hawn, entering her fifties, plays abandoned wives; and the postmenopausal Germaine Greer preaches that a woman's real sexual liberation is liberation *from* sex, in the life of a crone. Call it bravery or pathetic delusion, or both, only Mae West seems to have had a radically different notion and played it out for all to see. In 1954, at the age of sixty, she set off for Las Vegas.

In the Congo Room of the Sahara Hotel, she was carried onstage by four bruisers dressed in loincloths, and went through her old routines. She was aglow, in her element: she had not come down but back—to the only form of vaudeville that America had left. Reclining upstage on a divan, she assessed the endowments of a muscle-bound group who paraded on in capes they opened wide when they turned to face her. Sometimes she gave a little glance and kept on buffing her nails; sometimes she gasped; and when Mr. America did his pivot and turn, she got up and winked. She opened with "I Want to Do All Day What I Do All Night" and closed with "Frankie and Johnny." And she was a huge success. She claimed she was doing women a service—offering "the first bare-chest act for lady customers in history"—but women weren't much interested anymore. The semipornography was too blatant and too silly. Her most enthusiastic customers were now, more than ever, gay men, and she was unknowingly presiding over the second dawn of camp.

It is important to note that she was no longer fully in on the joke, or in on the biggest joke: her physical transformation and her blindness to it. This qualification distinguished the camp phenomenon that restored a crooked crown to her head in the 1970s from its larky thirties counterpart, and it could make the renewed sensibility seem uncomfortably cruel. Drag queens, now nearly feminist allies, continued to offer an object lesson in the artificiality of feminine accoutrements. But gay female idols were no longer the red-hot mamas, like Sophie Tucker, who had led up to West. Instead, they were the tragic, the stricken, the estrogenically doomed. By becoming an old woman—that is, by widening the gap between the mask of femininity and the reality behind it into a chasm that seemed as dizzying as if she were actually male—the strongest woman ever seen took her place among the new tragicomic goddesses with a dreadful, innocent pride in being applauded again.

Myra Breckinridge, West's 1970 return to the screen, marked the climax of a comeback that had ranged from a rock LP to a guest appearance on *Mister Ed.* Gore Vidal's transsexual satire was mangled to a point of toneless X-rated incomprehensibility: Vidal declared it the first time a movie had entirely halted the sales of a book. Unlike the bloody events of West's long-ago moralizing *The Pleasure Man,* the castration scene this time was self-chosen and surgical; it was only the movie that died. The director, Michael Sarnoff, announced from the start that he envisioned his faded star—West had a supporting role but top billing—as a drag queen. She continued to believe, however, that she was striking a blow for women. When the story had her septuagenarian character put in a hospital after a night with a young stud, West changed the script: "In my version, I put *him* in the hospital. See what I mean?"

It wasn't over yet. Her final film, the all but posthumous *Sextette,* concerned the sixth marriage of an ageless sex goddess (who happened to be eighty-three) to a twenty-something future James Bond (Timothy Dalton). Unable by now to remember lines or stage directions, the star had to be prompted from a radio transmitter hidden in her wig. She was tottery and got lost on the set. The film cost nearly eight million dollars and sat on the shelf for a year.

It was eventually shown at a San Francisco theatre in late 1978, on a double bill with Craig Russell's *Outrageous!,* a documentary about the artistic evolution of a drag queen—Russell himself, a former teen president of the Mae West Fan Club, who as a mere princess out to learn his trade had gained the star's support, and who paid her ample homage via a supreme Mae West imitation in his film. West's own film got second billing. She attended the premiere and refused to speak to Russell. In a review of *Sextette* for the *Times,* Vincent Canby wrote that West resembled "a plump sheep that's been stood on its hind legs," and advised that "Granny should have her mouth washed out with soap, along with her teeth."

Sextette has made a contribution to American scholarship, if nothing else, by at last committing to celluloid one of West's most famous lines, which had been mysteriously untraceable: "Is that a gun in your pocket or are you just glad to see me?" The shifting identity of the pocketed object during decades when the line floated free through the gender-riven American consciousness has been fully explored by one of the "media scholars" from our former English departments who have been producing a new trove of Westiana. "In conversation, a number of my female acquaintances have insisted that 'banana' belongs in place of 'gun,' " Ramona Curry, of the University of Illinois, reports in her book *Too Much of a Good Thing: Mae West as Cultural Icon.* Another of Professor Curry's acquaintances "recognizes the 'pickle' version as correct," however, and the advantages of both are indisputable: "The squishy edible, in place of the hard steel weapon, makes the bulging object palatable even as it deprives it of any power to wound."

If Susan Sontag's famed definition of camp as "failed seriousness" continues to serve, then American academic writing has taken over from old movies and old drag queens as our culture's leading camp phenomenon. The attraction of European critics to American popular arts and myths was based on their sense of renewal in encountering the frank, energetic pleasures of a national childhood not their own. Many American academics who currently write about our culture seem to have known nothing except this unending childhood, however, and to have only this culture to measure its products by—and they tend to write with a seriousness in inverse proportion to that of their subject. The new field of "star studies" depends on the examination of "star texts" with the aid of "fan discourse." Tools of critical analysis are brought to bear on episodes of *Mister Ed.* Of one such episode, Professor Curry notes that the appearance of Mae West "involves the program regulars, especially Mister Ed and Wilbur, in suggestively transgressive gender play," which develops even into "hints of bestiality."

The new enthusiasts see West as the ancestress of strong modern women: like Roseanne, but before the public confessions of childhood wounds; and like Madonna, but before the hard body and the baby. Their very changeableness, though, places these later pop icons in a different category. Nothing about Mae West ever changed. Part of that changelessness was being a good vaudevillian: you found what worked and you stuck to it. A lot of it was believing her act, and a lot more of it was actually being her act. It may never have been possible to tell the difference. There have always been those who have seen her as a secret victim, operating out of a kind of inverted weakness. Biographers have long speculated on her "lethal insecurity" (Cary Grant said he thought this was why she used so much makeup) and her use of sex as "the mighty panacea with which she remedied her life." It was without doubt the result of an extraordinary strength, from whatever source, that she managed to leave so little human evidence behind; to have a psychology at all is to court weakness, and so she did without. Three of her best-known plays of the twenties have recently been published—*Sex, The Drag,* and *The Pleasure Man*—and they reveal about all we'll ever see of the rough emotional edges that she planed into her Hollywood curves, and the head-on moral attack that she transformed into smiling subversion.

Fully conscious of her role in the battle of the sexes, and deservedly proud, Mae West wrote in 1959: "I did not perhaps treat the subject as seriously as Havelock Ellis, or as deeply as Sigmund Freud, Adler, Jung or Dr. Kinsey, but I think if we all could have sat down and discussed the subject fully, my ideas would have been listened to with some sense of awe. They may have been the generals, but I was in the front lines—out in an emotional No Man's Land, engaged in dangerous hand-to-hand, lip-to-lip raiding parties." Maybe the best of her valor and the choicest of her victories—for both armies—arose from how much she enjoyed the fight, and insisted on her right to enjoy it and, of course, on our right to enjoy her enjoying it. What it cost her we'll never know. When, in 1980, Mae West had a stroke—she died a few months later—she told the papers she'd fallen out of bed because she'd been dreaming of Burt Reynolds; and it had been worth it.

A Study in Scarlett

Margaret Mitchell

In the summer of 1936, American literature divided resoundingly along its oldest fault line, and the resulting chasm seemed to grow wider and deeper with every sale—roughly a million by the end of December—of a fat new novel called *Gone with the Wind.* On one side of the break, patently serious writers and critics conceded that they were hopelessly outnumbered—a fact that the representatives of literature had been bemoaning since at least the middle of the previous century, even before Nathaniel Hawthorne's famous complaint that he was being driven from the literary marketplace by "a damned mob of scribbling women" and a public taste "occupied with their trash." It was one such scribbler and her public who now thronged the opposite side.

The fear of a downwardly spiralling culture associated with a new mass audience had taken on, in literature, the specific taint of the superficial sex. While men who did not understand literary art could be counted on, for the most part, to stick to newspapers, the lettered (if not highly educated) female population had long monopolized sales of fiction, corrupting the novel from its noble roots in Romance—in the greater historical sense, as a worldly or spiritual quest—into romance in the distinctly lesser sense of a courtship tale culminating in marriage. In 1852 Harriet Beecher Stowe produced a novel that was as decried for its domestic bathos as it was celebrated for its moral influence, and that went on to become the biggest best-seller the United States had yet known. (Hawthorne's complaint followed *Uncle Tom's Cabin* by less than three years.) This lengthy precedent could be felt as a kind of pressure slowly building toward the rending contradictions of Margaret Mitchell's Civil War extravaganza: a triple-decker Victorian romance issued nearly twenty years after the Joycean disruptions of modernism; a book by an unknown writer that sold more copies in its first few weeks than many major authors sold in their lifetimes; a story that took root in the national imagination with the rampant force of a myth or a psychosis; America's favorite novel and no part of its literature.

Staggered by the sales figures, distinguished critics were reduced to assailing the patrons of bookstores for being far too eager to reach into their pockets—or, rather, their purses, since, it was recalled, "most book buyers are women"—to pay the unheard-of price of three dollars for what Malcolm Cowley, efficiently summing up both book and audience, characterized as an "entertainment that will carry them through the idle moments for a whole fortnight." One of the striking things about initial critical reactions to Mitchell's work, for and against, was their absolute accord over what it offered—powerful storytelling—and what it lacked: literary style and originality. It was in the value placed on these apparently opposed qualities that ways parted and stands were taken, and the question of whether the term "popular literature" could ever again signify anything more than a bitter oxymoronic joke was widely if sometimes implicitly argued.

Mitchell's book was continually praised for its "readability," as though this were not the first and simplest requirement of any book. For a vast audience, however, the logic of this basic proposition had collapsed some years before. In October 1936, when William Faulkner published a very different story of the South and the causes and effects of the war, *Absalom, Absalom!*, the *Times*, in a review typical of those the book received, credited it with "one of the most complex, unreadable and uncommunicative prose styles ever to find its way into print." Like *The Sound and the Fury* and its other predecessors, Faulkner's new work won only occasional, if intensely felt, praise—for its moral vision, and for what Mitchell's home bastion, the *Atlanta Journal*, recognized as the "first real step forward" in the novel form "since *Remembrance of Things Past*." Fourteen years before Faulkner was awarded the Nobel Prize in Literature, fifteen years short of the work's reissue as a Modern Library classic, *Absalom, Absalom!* sold about seven thousand copies and then disappeared from the shelves.

Despite Mitchell's evident victory, there was the tone of a counteroffensive in those critical celebrations which crowed over the way she "tosses out of the window all the thousands of technical tricks our novelists have been playing with for the past twenty years," and which emphasized the importance of her book as "an alternative to the pessimism, obscurity and fatal complexity of most contemporary novelists." These commendations were offered by Herschel Brickell of the *New York Post* and Edwin Granberry of the *New York Sun*, two of Mitchell's fervent champions, and Granberry concluded, point-blank, "Could it be possible that *Gone with the Wind* might make it difficult hereafter for the pinched, strangulated novel which pays more attention to manner than matter?"

Even Cowley, one of Mitchell's harshest early critics, found it possible to conclude that, while *Gone with the Wind* was indubitably not a great novel, it did, almost incredibly, make us "weep at a deathbed (and really weep)" and "exult at a sudden rescue," and that it possessed "a simple-minded courage that suggests the great novelists of the past." In fact, among Mitchell's boldest advocates, both *War and Peace* and *Vanity Fair* were frequently evoked in assessments of her novel's historical scope and its contrasting pair of leading ladies. (An Atlanta librarian gave a speech introducing Mitchell in which, quite astutely, she added *Gentlemen Prefer Blondes* to the venerable list.) Small wonder, then, that in the excitement of its arrival *Gone with the Wind*, seen by some as the last popular straw, was viewed by others as an exemplary way out of an ever-narrowing and more exclusive modernist dictate, a reading ground of reconciliation for a democracy's divided audience.

The book moved into an even larger realm of democratic access when it was sold to the movies, a month after publication, for the record sum of fifty thousand dollars. Although the film, produced by David O. Selznick, was the subject of intense national curiosity throughout the three years it was being made, Mitchell refused to have anything to do with it, apart from recommending Georgia friends to serve as consultants on matters of authenticity in custom, dress, and even Southern horticulture. (They kept dogwoods from blooming during cotton-picking time, and the cotton itself from springing up along a plantation's front lawn.) Mitchell herself never set foot in Hollywood.

Arriving there, however, at just the time of the big sale was William Faulkner, forced to hire himself out as a screenwriter—not for the first time—after the commercial failure of all his recent work. A letter written that September suggests his reaction to the *Gone with the Wind* phenomenon: shifting among film assignments like *Slave Ship* and *Splinter Fleet,* he announced to his agent that he was determined to sell *Absalom, Absalom!* to the studios himself, and furthermore, he said, "I am going to ask one hundred thousand dollars for it or nothing." Nothing is what he got. Faulkner made it clear that he had not read Mitchell's book ("No story takes a thousand pages to tell" was his full pronouncement), and his only other reflection on its significance may be inferred from a letter written in the summer of 1936, from Mississippi, in which, updating Hawthorne, he lamented his lost habit of "writing trash" and added, "I seem to be so out of touch with the Kotex Age here."

By late 1939, when Selznick's *Gone With the Wind* was nearing release, a Gallup poll found that an estimated 56,500,000 people were planning to see it. Fulfilling all expectations, the overwhelming adulation accorded the movie and its stars served to reinforce the popularity of Mitchell's book, and has very possibly preserved it. In the public mind, the two versions have merged to the point where it is difficult to say anymore whether *Gone with the Wind* is in essence a novel or a movie, and, in fact, the distinction may not mean much: Mitchell's characters long ago burst through the restraints of their form and, like folk- or fairy-tale figures, passed directly into the mainstream consciousness.

The much remarked "readability" of the book must have played a part in this smooth passage from the page to the screen, since "readability" has to do not only with freedom from obscurity but, paradoxically, with freedom from the actual sensation of reading—of the tug and traction of words as they move thoughts into place in the mind. Requiring, in fact, the least reading, the most "readable" book allows its characters to slip easily through nets of words and into other forms. Popular art has been well defined by just this effortless movement from medium to medium, which is carried out, as Leslie Fiedler observed in relation to *Uncle Tom's Cabin,* "without loss of intensity or alteration of meaning." Isabel Archer rises from the page only in the hanging garments of Henry James's prose, but Scarlett O'Hara is a free woman.

When, in 1913, Henry James saw one of the innumerable stage productions of *Uncle Tom's Cabin* that had sustained the book's fame and its message over the decades, he recognized this "leaping" quality of popular art. "Uncle Tom," he wrote, "instead of making even one of the cheap short cuts through the medium in which books breathe, even as fishes in water, went gaily roundabout it altogether, as if a fish, a wonderful 'leaping' fish, had simply flown through the air." Having accomplished this feat, "the surprising creature could naturally fly anywhere, and one of the first things it did was thus to flutter down on every stage, literally without exception, in America and Europe." Margaret Mitchell labored over a book, not a screenplay; she doubted for a time whether *Gone with the Wind* could be filmed at all. ("I don't see how it could possibly be made into a movie," she wrote to her publisher, who had intimated otherwise, "unless the entire book was scrapped and Shirley Temple cast as 'Bonnie,' Mae West as 'Belle,' and Stepin Fetchit as 'Uncle Peter.' ") What her work reflects to perfection is the state of affairs in the republic of letters during a period when all popular art aspired to the condition of the movies.

Yet the ancestry of *Gone with the Wind* extends back far beyond Hollywood. In a personal genealogy sketched out for her publisher, Margaret Mitchell noted that some of her forebears had sailed to America "with the Hector MacDonald colony after the failure of the Stuart uprising," a family legend that constitutes only the most literal element in the author's heritage from Walter Scott. The valiant Scottish clans of the Waverley novels, aligned against the English in the cause of the exiled Stuart kings, were a near worldwide sensation in the early nineteenth century, and Scott's romantic nationalism stirred a deep chord of response from France to Italy to Russia—which is to say, from Balzac to Manzoni to Pushkin. In the raw new American literature, Sir Walter's twilight-of-a-nobility theme was reworked in books like James Fenimore Cooper's *The Last of the Mohicans,* in which the tragic aristocrat of the Highlands was resettled among the tribes of the New World's frontier. (Cooper begins with an epigraph from *Richard II:* "Say, is my kingdom lost?") But for all their renown, the works of Walter Scott, and the gospel truths to be found therein, were cherished nowhere else so long or so well as in the American South.

With Scott's *Ivanhoe*—issued among the Waverleys for a variation in setting, if not in theme—an idyll of sentimental feudalism was seriously taken up, in the antebellum South, as a blueprint and a benediction for a society already divided into landed fiefdoms and fully regulated by caste. Out of the novel's Arthurian bombast, the states of the future Confederacy fashioned an elaborately archaizing cult of courtliness (the leading planters dubbed themselves "the Chivalry"), complete with tournaments and duels and, above all, a highly exaggerated attachment to the chastity and honor of women, who were reared and cultivated accordingly. The adoption of this fantastic, mass-scale impersonation—and had there been no Scott, his defenders have pointed out, Malory would have done as well—served to transform the surface appearance of a brutal and retarded economic system into a fancy-dress theatrical. It also provided a much needed cultural ambience—based, of necessity, on the fullest amplification of social ceremony—in a region that, as European and Yankee visitors noted, was conspicuously lacking in other signs of culture, from orchestras and opera houses to publishing firms and libraries and debating societies, and in which censorship had severed access to all intellectual engagement with the larger issues of the political order.

By the time Mark Twain had steamed down the Mississippi as far as the neo-Gothic, turreted statehouse of Baton Rouge and the Mardi Gras in New Orleans, he felt able to pinpoint the source of all the errors and woes of the deluded and darkened—indeed, the anti-Enlightenment—South:

> Then comes Sir Walter Scott with his enchantments, and by his single might checks this wave of progress, and even turns it back; sets the world in love with dreams and phantoms . . . with the sillinesses and emptinesses, sham grandeurs, sham gauds, and sham chivalries of a brainless and worthless long-vanished society. He did measureless harm; more real and lasting harm, perhaps, than any other individual that ever wrote. . . . It was Sir Walter that made every gentleman in the South a Major or Colonel, or a General or a Judge, before the war; and it was he, also, that made these gentlemen value these bogus decorations. . . . Sir Walter had so large a hand in making Southern character, as it existed before the war, that he is in great measure responsible for the war.

It is plainly no accident that Huck and Jim are nearly done in by a gang of murderers aboard a wrecked steamboat called the *Walter Scott.*

"Say, is my kingdom lost?" would have been an appropriate epigraph for *Gone with the Wind,* as it would have been, too, for the half century of "plantation novels" that preceded it. The type emerged full-bodied and heavy-scented in the 1880s, drenched in nostalgia for the way of life the war had taken, although, as the historian William R. Taylor pointed out more than two decades after the apotheosis of *Gone with the Wind,* the nostalgia actually predated the war. Before the "antebellum" was ante anything, it was merely an aftermath, from which Southerners who were so inclined yearned for the brighter paradise before the Revolution. Southern novelists from the 1830s on harked back to the ancien régime of Colonial Virginia with moonstruck longing, as in John Esten Cooke's *The Virginia Comedians* of 1854, quoted in Edmund Wilson's *Patriotic Gore:* "Where are they now, those stalwart cavalries and lovely dames who filled that former time with so much light, and merriment, and joyous laughter? . . . What do we care if the laces are moth-eaten—the cocked hats hung up in the halls of Lethe—the silk stockings laid away in the drawer of oblivion?" Of course, yearning for a lost golden age may be less a response to a real historical place and circumstance than a chronic human inclination, and perhaps a precondition of the literary impulse; even Homer had to look back centuries to find heroes worthy of his praise.

In the 1880s, the South's premier plantation novelist, Thomas Nelson Page, invented or codified every cliché of worthy master and loyal slave, and his works were popular not only in his home region but in the repentant and conciliatory North. After the war, of course, it became as safe for Northerners to vent a retrospective sigh for the age of Massa and Mammy as it had been for Walter Scott to exalt the ancient glamour of the Stuarts from a position securely founded on Whiggish prosperity.

Its image enhanced by the ineffable charm of loss—of having lost, of being lost—the South assumed its role in the romance of America as the festooned and feminine counterpart of the relentlessly masculine West. So appealing did the image of languid Southern gentility come to seem in an age of unstoppable industrial momentum that even W. E. B. Du Bois could write with lyrical regret of the passing of "the old ideal of the Southern gentleman,—that new-world heir of the grace and courtliness of patrician, knight, and noble." In his eulogizing essay "Of the Wings of Atalanta," collected in *The Souls of Black Folk* in 1903, Du Bois lamented the South's accession to a greedy new mercantile culture and compared his adopted city of Atlanta, symbol of all that the South might be, to the legendary Greek girl ("If Atlanta be not named for Atalanta, she ought to have been") who outraced all men but lost her freedom when tricked by gold.

Margaret Mitchell, an Atlantan of six generations' standing, claimed as one of her novel's first aims the creation of a heroine who would embody the vital contradictions of the South's most ambitious city—a city "crude with the crudities of youth and as headstrong and impetuous as herself." The fact that she even attempted this was testimony to her larger claim to have broken with the old plantation tradition. Northern critics like Cowley might see her book as "an encyclopedia of the plantation legend," and Louis Kronenberger, writing in *The New Yorker,* could imagine Mitchell waking in the night to groan, "I left out a lynching! I left out a fox hunt!"; but Southerners like Stephen Vincent Benét registered the book's differences from its predecessors and praised its author's "more realistic treatment."

Mitchell fully believed in the daring of her realism. On the verge of the book's publication, her husband instructed the Macmillan Company to prepare for "a 'public relations' problem" that *Gone with the Wind* might arouse "in these parts"—quite reasonably, in his view, since "it deals with the South with considerable frankness." What he evidently had in mind was the fact that Mitchell's up-country Georgians are not old landed gentry but a socially mixed and rambunctious lot, many of them so newly settled that even an Irish immigrant—Gerald O'Hara, who had won his plantation, Tara, in a card game—could gain a place among them. O'Hara's neighbors, the horse-breeding and slave-owning Tarletons, have "less grammar than most of their poor Cracker neighbors"; the only "aristocrat" in their household is their dog. As for the patrician Wilkes family, of nearby Twelve Oaks—that "beautiful white-columned house that crowned the hill like a Greek temple," without which a Southern tale would have been like a mystery without a crime—they are thought by their upstart neighbors to be "born queer," partly because of inbreeding but mostly, as Ma Tarleton tells her boys, "because their grandfather came from Virginia."

The historian Henry Steele Commager observed, in an early and highly favorable review, that *Gone with the Wind* was indeed about the opposition of two civilizations, but that these were not the North and South but the Old South and the New. The momentous giving way of one era to the next is traced by Mitchell over a span of twelve years, from the eve of the Civil War to the middle of Reconstruction. But the discord and instability of the time are dramatized, above all, in the person of a girl, introduced on page one at the age of sixteen, whose very face betrays the contrast, too sharp, of "the delicate features of her mother, a Coast aristocrat of French descent, and the heavy ones of her florid Irish father"—a fast and greedy young Atalanta named Scarlett O'Hara.

▪ ▪ ▪

Margaret Mitchell professed annoyance when asked, as she often was—and at least once for good money, by *Vogue*—to discuss the character of Scarlett O'Hara in terms of her "modernity." "Good God," she remarked in a letter that first hectic September, "do they think hard-headed women only came to life in the 1930's? Why don't they read the Old Testament?" But the progress of the character was confounding even to the author. Shortly before publication, Mitchell had requested that the name of Scarlett's kindhearted foil, Melanie, not be removed from the advertising copy—she is still there, described on the current dust jacket as "a loyal friend and true gentlewoman"—because, Mitchell avowed, "after all, she's the heroine of the book." (Thackeray had affixed the same label to the "gentle and uncomplaining little martyr" Amelia Sedley midway through *Vanity Fair*, in pointed distinction to another such green-eyed baggage as Mitchell's Scarlett.)

Scarlett O'Hara became enough of a public obsession and seemed enough of a credible personality to be psychoanalyzed in learned journals, and to a psychiatrist who concluded that America's new princess was a "partial psychopath" and a person of "inward hollowness," Mitchell responded with excited approval; at last, someone had got her point. "I set out to depict a far-from-admirable woman," she wrote in one of the long letters that consumed her time in the book's aftermath, and which have been collected and edited by Richard Harwell. "I have not found it wryly amusing when Miss O'Hara became somewhat of a national heroine and I have thought it looked bad for the moral and mental attitude of a nation." True, Mitchell could also, on occasion, defend her character's finer qualities of courage and perseverance and appetite for life. One popular diagnosis of just why *Gone with the Wind* had conquered America—it appeared in the *Reader's Digest* in 1939—found that the story's primary appeal lay in Scarlett O'Hara's position as "the master of her world rather than its victim," her exemplification of "personal triumph over social insecurity." It may indicate something about changing times that in 1957 a survey of a class of American high school girls, noted by Helen Taylor in the book *Scarlett's Women*, found that all but one of the girls identified with docile Melanie, or claimed to; and that in 1970, in another survey mentioned by Taylor, three quarters of the girls firmly aligned themselves with Scarlett.

Of course, no one needed to ask whether young American males identified with the book's fair Ashley Wilkes or dark Rhett Butler, for it was reasonably certain that few had read it. Despite its wartime setting, *Gone with the Wind* is in no respect a "war" book: the gallant soldiers leap onto their horses and ride off to battle as if over the edge of the earth, and they return or they don't; the reader's place is emphatically with those who stay behind. All is seen from a woman's point of view—or, rather, from a girl's. The tempting young men are nearly overmatched by the tempting dresses, the "rose organdie with the long pink sash," or the "green plaid taffeta, frothing with flounces and each flounce edged in green velvet ribbon," or the "butter-yellow watered silks with garlands of rosebuds." In the long period during which Mitchell's audience of American "girls" could retain their status until late middle age, and then on through the years of feminist tension between growing freedom and obdurate complicity, Scarlett O'Hara has remained a congenial paragon of contradiction: a prodigy of femininity in full rebellion, an expert in the disdained tactics of sex, a "master of her world" who is never less than wildly desirable—capability and authority with a seventeen-inch waist.

Mitchell's favorite word for Scarlett is "unanalytical," but from the start the girl is aware that even the most cultivated bloom of ingenuous charm cannot make her into the lady she aspires to be. Unlike Becky Sharp, Scarlett is troubled by her failure of gentility, and troubled, too, by her ability to see through the mechanisms necessary to deliver her to her fate: "Don't you suppose men get surprised after they're married to find that their wives do have sense?" It hardly needs to be pointed out that the Southern belle was bred to conform to a subspecies of the nineteenth-century "lady" that exceeded all other varieties in its veneration of a high artificiality, in its observance of a prescribed distance between ideal and flesh. "At no time, before or since, had so low a premium been placed on feminine naturalness," Mitchell writes, and in this historical judgment, at least, she stands confirmed. For Scarlett, however, the ideal takes an entirely natural form in her adored mother, the saintly Ellen, whose back is never seen to rest against the back of any chair on which she sits, whose broken spirit is everywhere mistaken for righteous calm, and in whose chaste perceptions—despite her three daughters and three buried sons—"mares never foaled nor cows calved," and even "hens almost didn't lay eggs." Scarlett hopes that someday she will manage to be like her mother, only—rather like St. Augustine—not yet.

But why should so extreme an attitude toward women—"gyneolatry," as one historian has termed it—have prevailed in the American South? Is it to be understood simply as a fevered symptom of the Walter Scott disease? Or was it a deeper response—not to the neochivalric delusion but to the society's need to be deluded? Margaret Mitchell troubled her head with "why" no more than Scarlett does. But then this is a question to which Mitchell could not possibly attend, since the answer, like so many answers in the South, is tied to the facts of slavery and race, facts that the author cannot abide and that she spent considerable energy in dissembling.

The only antebellum chains glimpsed in *Gone with the Wind* are metaphorically attached to the hardworking ladies of the manor houses—"chained to supervision of cooking, nursing, sewing and laundering." It has become a commonplace observation that Harriet Beecher Stowe approached the unfamiliar oppression of slavery through an intimate knowledge of the oppression of her sex. Charlotte Brontë's comment that "Mrs. Stowe had felt the iron of slavery enter into her heart from childhood upwards" is roughly contemporary with the remark of Mary Chesnut—no friend to Mrs. Stowe—that "there is no slave, after all, like a wife." It was through this kind of identification with enslavement that many middle-class white women came to the forefront of the abolitionist movement, and then went on to establish the cause of their own suffrage with a sense of natural progress and undeniable justice. The slave narrative of Harriet Jacobs approaches its conclusion with the perfect reversal of the Brontë romantic formula: "Reader, my story ends with freedom; not in the usual way, with marriage."

In the South, the connection between women and slaves, no less strong, came to be expressed not as an alliance but—particularly after the rise of abolitionism—as a necessary opposition. Slavery was represented, by its supporters, as a specific boon to women. An 1832 study which included an account of the last great debates on slavery held in the Virginia legislature argued that the institution served gloriously to lift "woman" to a new and fitting station: "We behold the marked effects of slavery on the conditions of woman—we find her at once elevated, clothed with all her charms, mingling with and directing the society to which she belongs, no longer the slave but the equal and the idol of man."

But there is a more insidious element in this elevation of a new idol than can be explained by the transfer of labor from wife to slave, and it is an element that Margaret Mitchell delicately but quite certainly meant to address, or, rather, to dismiss, by her repeated descriptions of the admirable slaves of Tara. Both Mammy, "shining black, pure African," and Pork, "shining black, dignified," are pointedly of unmixed blood (only Dilcey is said to be part Indian)—a characteristic that distinguishes them from the substantial number of mulattoes, who by the 1860 census accounted for a minimum of 12 percent of the nonwhite Southern rural population. In the contemporary words, again, of Mary Chesnut, who was a frequent resident of her father-in-law's South Carolina plantation and a friend of the Jefferson Davises: "We live surrounded by prostitutes. . . . Our men live all in one house with their wives and their concubines, and the mulattoes one sees in every family exactly resemble the white children—and every lady tells you who is the father of all the mulatto children in everybody's household, but those in her own she seems to think drop from the clouds, or pretends so to think." And, elsewhere, "Mrs. Stowe did not hit the sorest spot. She makes Legree a bachelor."

Looking at passages from Chesnut's diary and similar accounts, Edmund Wilson felt forced to conclude, as W. J. Cash had done two decades before him, that "the pedestalled purity which the Southerners assigned to their ladies, the shrinking of these ladies themselves from any suggestion of freedom, were partly a 'polarization' produced by the uninhibited ease with which their men could go to bed with the black girls." To set oneself off from the status of mere "chief slave of the harem," in the words of a planter's wife recorded by Harriet Martineau in 1837, one would have to set oneself off from the flesh itself. To be above reproach was also, perhaps, to be above feeling reproach; to think no evil was to see none. Her moral virginity intact, her sanction granted, the Southern woman's status—"the South's Palladium," Cash called her—was her reward. There was far more to Ellen O'Hara's immaculate cows and hens than Margaret Mitchell would ever be willing to admit.

Mitchell's mother, Maybelle Mitchell, was indisputably a lady, and something more: around the time of her daughter's birth in 1900, she became one of Atlanta's leading suffragists. This radical cause would seem to contradict the antebellum sentiment with which the family also lived (Margaret's older brother was named for the vice president of the Confederacy, Alexander Stephens), the natural residue of a culture still bitter over defeat and essentially unresigned; in those years, Atlanta displayed one United States flag—at the post office. The twin dedications of the household were reconciled, evidently, in the conviction that women's strength was a foundation of the Old South. (Living proof was ever at hand in the dauntless person of Maybelle's own mother, Annie Fitzgerald Stephens, whose legendary trials and triumphs throughout the war were to serve as inspiration for those of Scarlett O'Hara.) And so the little girl who grew up singing "I'm a Good Old Rebel" as a parlor trick was also carted off to suffrage rallies with a "Votes for Women" banner tied around her belly.

Maybelle Mitchell had forfeited hopes for a career in science or medicine when she married, and she was driven by the desire that her daughter acquire an education. Margaret Mitchell later recounted payment schemes by which her mother coerced her into "classical" reading: "Mother used to give me a nickel for each of Shakespeare's plays . . . a dime for Dickens, fifteen cents for Nietzsche and Kant and Darwin." It had not been an easy way to augment her allowance: "Even when she raised the ante to twenty-five cents with a licking thrown in, I couldn't read Tolstoy, or Hardy or Thackeray either, for that matter." By the time she repeated this story, in letters to appreciative fans and reviewers, Mitchell was easing into her role of Good Ol' Girl of the best-seller lists, explaining why her Scarlett could not have been indebted to Becky Sharp: she hadn't read *Vanity Fair* until 1935. (This was, perhaps coincidentally, the year of the RKO Technicolor *Becky Sharp,* starring Miriam Hopkins, whom Mitchell privately touted for the role of Scarlett.)

Another of Maybelle's attempts to impress her principles on her daughter's mind resulted in an experience that Margaret Mitchell called "the genesis of my book"—her first unforgettable lesson in what she saw as the great theme of survival. Aged six, Margaret had returned from the new experience of school angry and discouraged, refusing to return. "And Mother took me out on the hottest September day I ever saw," she recounted in 1936, "and drove me down the road toward Jonesboro . . . and showed me the old ruins of houses where fine and wealthy people had once lived. . . . And she talked about the world those people had lived in, such a secure world, and how it had exploded beneath them. And she told me that my own world was going to explode under me, some day. . . . She said that all that would be left after a world ended would be what you could do with your hands and what you had in your head."

The Jonesboro road that mother and daughter took that day became, Mitchell attests, "the road to Tara," along which Scarlett O'Hara would flee from burning Atlanta in her rickety wagon through the charred and empty countryside. "If she could only reach the kind arms of Tara and Ellen," Mitchell wrote, "and lay down her burdens, far too heavy for her young shoulders—the dying woman, the fading baby, her own hungry little boy, the frightened negro, all looking to her for strength, for guidance, all reading in her straight back courage she did not possess and strength which had long since failed." Scarlett arrives home, the house still standing but emptied of comfort or rest, on the day after her mother's death.

Margaret Mitchell said that she wrote this part of the book, and only this part, in one long streak: twenty-four pages without revision. Critics generally found it to be the best of her work. This is a matter not of finer sentences or more elegant phrases—Mitchell's effects are not to be found in such separable elements—but of the broad, building rhythms sustained over the journey, the discovery, and the final resolution. Her world overthrown, Scarlett takes up the new matter of hunger, her own and her family's, and when weariness and illness overcome her she makes the vow that meant so much to Mitchell's barely post-Depression audience: "I'm going to live through this, and when it's over, I'm never going to be hungry again. No, nor any of my folks. If I have to steal or kill—as God is my witness, I'm never going to be hungry again." And yet, for all that Maybelle Mitchell's lesson struck deep, her real point was lost in the mounting violins and the lurid sky, even in the book. Her final plea, after all—the reason for the ride into the ruined past—was, as always, the importance of education. "So for God's sake, go to school and learn something that will stay with you" is what her daughter remembered her saying when their journey was over.

In fact, Margaret Mitchell not only resisted her mother's attempts to give her a literary education but mutinied completely around the age of twelve, tumbling free at last into pulp and adventure. Movie-crazed, she grew into a determined madcap, a four-foot-ten-inch "baby-faced li'l vamp"—an overheard description that delighted her—honing her skills on the soldiers of a nearby military camp. In the fall of 1918, she went off to Smith College, where she received mediocre grades and an enviable quantity of mail from servicemen overseas. In what seems now the outstanding incident of her college career, Mitchell quit a history course, in anger, because a black student was also enrolled; in defiance of college rules, she managed to obtain a transfer to another class. A more apt allegory for Margaret Mitchell's relationship to the subject of African-Americans in history could hardly have been invented.

In the middle of her freshman year, Mitchell was summoned back to Atlanta by news that her mother was ill. Maybelle Mitchell died of the flu in January 1919, on the day—like Ellen O'Hara—before her daughter reached home. Foreseeing the consequences of her death, she left Margaret a letter that fairly glows with maternal light, counselling her against the temptation to surrender her future in order to become her father's caretaker: "Give of yourself with both hands and overflowing heart, but give only the excess after you have lived your own life." Margaret returned to Smith to finish out the year, and then came home for good, to do exactly what her mother had warned her against. In later years, she made grandiose claims for the career she had sacrificed—"I started out to be a psychiatrist, but, unfortunately, was forced to leave college when my mother died"—but it seems clear that what she experienced at the time was a sense of relief.

Back in Atlanta, Mitchell became the classic jazz baby—by her own description, "one of those short-haired, short-skirted, hard-boiled young women who preachers said would go to hell or be hanged before they were thirty." She scandalized the Junior League by performing, at a charity ball, an Apache dance—complete with clinch—adapted from a Valentino movie. The men from nearby Camp Gordon met each other coming and going on her Peachtree Street veranda, and at one point she was engaged to five of them. "You can say all you please about my being an unscrupulous flirt," she wrote to a college friend in the North, "but I'm here to state that I haven't lied to those five men—nor have I misled them in any way."

In 1922 she made a disastrous but quickly ended marriage to a handsome bootlegger who had been forced to resign—twice—from Annapolis. She got a job writing features for the Sunday magazine of the *Atlanta Journal*—reporter jobs were not open to women—and she sent off a group of stories to H. L. Mencken's *The Smart Set:* all were rejected. (Mencken had recently published his notorious essay on the South, "The Sahara of the Bozart," in which he identified even the once reigning state of Virginia as "an intellectual Gobi or Lapland" and characterized the whole region as "a vast plain of mediocrity, stupidity, lethargy, almost of dead silence.")

In July 1925 Mitchell was married again—to John Marsh, a suitor who had lost out the first time and had stayed on to serve as best man at her previous wedding. Marsh was as mild and conventional as her first husband had been wild. Having asked her father's consent and set the date, he became ill and had to be hospitalized with severe hiccups, an ailment that persisted for forty-two days. After the marriage, Mitchell continued at her newspaper job for nearly a year, until Marsh received a raise from the utilities company where he worked, and, bowing to his wishes, she agreed to stay home. But it was while she was still at the *Journal* that she began, with

Marsh's encouragement, to work on what he would later refer to as a "Jazz Age novel." Marsh had given her a copy of *The Great Gatsby,* newly published, for their first Christmas together. She was already an admirer of Fitzgerald's work, and she described him years later—in 1939—as a kind of hero, even as Fitgerald himself was toiling away in Hollywood helping to pare her dialogue into a screenplay.

The best bits of *Gone with the Wind* do indeed reflect this literary taste and this initial ambition. (That Scarlett was originally to be called "Pansy" may also suggest the incentive of Fitzgerald's Daisy.) They have a lightness and a barely transplanted "Jazz Age" fizziness that manage to lift the text briefly on little gusts of social comedy: the Tarleton twins' scheming to fall in love with the same girl so as to keep each other company; the crossed signals of Charles Hamilton's warm proposal of love and Scarlett's frosty response; some of the descriptions of Southern customs ("Frequently elderly aunts and uncles came to Sunday dinner and remained until they were buried years later"). The princely Ashley Wilkes speaks as though his mind were in a truss, but to Scarlett he is ever, quite simply, "the tall drowsy boy she loved."

Phrases that ring clear have most often to do with Scarlett's character—determined and gay, childlike and utterly selfish—as she finds herself, confused and resentful, in acts of necessary kindness: yielding her bonnet to shield Melanie from the beating sun on the road to Tara, her only thought is the endearingly grumpy "I'll be as freckled as a guinea egg before this day is over." And in what must be the sweetest line in the book, Scarlett ministers, unhappily, to the broken young Confederate soldiers as they retreat through Atlanta: "Why should she be standing here in Aunt Pitty's peaceful front yard, amid wavering lights, pouring water over dying beaux?" Those critics and readers who vexed Mitchell by insisting on her heroine's "modernity" may have been on to something of the Jazz Age girl still lurking deep within Scarlett O'Hara's vast hoop skirts.

Such qualities of gaiety or sweetness as *Gone with the Wind* possesses are confined largely to the book's earlier sections, before the onset of Reconstruction or the felt consequences of Emancipation. ("It's just ruined the darkies," says Scarlett, innocent of irony, if of little else.) But even from the start these patches of light are heavily overshadowed, and are finally blotted out entirely by the inescapable grimness of Mitchell's racial politics.

Approximately three quarters of the way through the novel, Scarlett

O'Hara is assaulted while riding alone in her carriage. The war is over. Her first husband, the calflike Charles Hamilton, whom she married for spite, left her a widow. Her second husband, the meek Frank Kennedy, whom she married for money, has been unable to prevent her, despite the breach of propriety and the danger, from pursuing her business interests in outlying parts of Atlanta. It is almost dark, and she is on the road bordering the new postwar Shantytown, when she is set upon by "a big ragged white man and a squat black negro with shoulders and chest like a gorilla." She refuses to give them money; the white man shouts that it must be hidden "in her bosom." Her gun is wrenched from her hand, and then: "The negro was beside her, so close that she could smell the rank odor of him as he tried to drag her over the buggy side. With her one free hand she fought madly, clawing at his face, and then she felt his big hand at her throat and, with a ripping noise, her basque was torn open from neck to waist. Then the black hand fumbled between her breasts, and terror and revulsion such as she had never known came over her and she screamed like an insane woman."

Scarlett is saved at the last moment by the appearance of Big Sam, the onetime slave foreman of Tara. He beats both men off—perhaps kills them. ("Ah hope Ah done kill dat black baboon. But Ah din' wait ter fine out," he tells her afterward. "But ef he hahmed you, Miss Scarlett, Ah'll go back an' mek sho of it.") Sam drives her home and is thanked for his loyalty. And that night Scarlett becomes a widow again, when her husband is killed during a retaliatory raid on Shantytown carried out by the gallant white knights of the Ku Klux Klan.

Elements of this lengthy episode had been in common use for years when Mitchell came to write it. She had constructed it as an alternative version, deliberately eventful, after one in which Frank Kennedy died of illness seemed to her too dull. Scarlett's husband had to be got rid of somehow, and Mitchell submitted the alternative deaths with her manuscript. An outside reader hired by Macmillan to appraise the work—Charles Everett, a Columbia English professor, who was, over all, highly enthusiastic—advised that the author forgo the Klan episode "because," he diplomatically suggested, "the KKK material has been worked pretty hard by others." Mitchell disclaimed any special political regard for this set of events; her preference was owed only to its greater liveliness. "As 'Alice' would have said," she explained in what might well pass for her literary credo, there would otherwise have been "no conversation and absolutely no pictures" in that part of the book.

This particular line of conversation and these particular pictures—the fragile white woman clawed at by the black savage and avenged by the knights in white sheets—had been in circulation ever since the Klan itself was organized and began to grow, just after the war. The opposition the South had asserted between the position of its slaves and that of its women as the balancing forces of Confederate theology—Devil and Virgin, sin and forgiveness, the damned and the exalted—turned the spectre of rape into a compelling blasphemy; in a society long quietly familiar with illicit sex based on ownership, the possibility of revenge was so tensely awaited as to become a fixation. Scarlett O'Hara's unlucky carriage ride is carefully dated to a windy March day during "military" Reconstruction, soon after "the legislature refused to ratify the amendment"—the Fourteenth—and during a period when, the author informs us, "it was the large number of outrages on women and the ever-present fear for the safety of their wives and daughters that drove Southern men to cold and trembling fury and caused the Ku Klux Klan to spring up overnight."

In reality, the cry of rape as a political weapon was only beginning to be heard in the years after the war. It took its place on a long list of crimes to be avenged by the vigilantes of the Klan, or by the Knights of the White Camellia or any of the other more fleetingly established fraternities of terror—crimes like voting (either for the wrong party or at all), renting the wrong land, working at the wrong job, and in general being seen to behave in a manner suggesting the serious nature of Emancipation (a manner known in common parlance, and in Mitchell's, as being "uppity," as in Aunt Pitty's "The Yankees are very upset because so many uppity darkies have been killed recently"). It was with the calamitous agrarian depression of the 1890s, when the Northern pressure for Negro rights abated and the South began to hope for disenfranchisement and to gather evidence for its necessity, that rape suddenly became an obsession, the overwhelming threat of the historical moment projected with equal conviction into the future and onto the past. By the time Margaret Mitchell was old enough to read, the history through which her heroine would ride had been largely rewritten.

"I was practically raised on your books, and love them very much," Mitchell wrote in answer to a fan letter from the novelist Thomas Dixon in 1936. This gracious and rather effusive tone was to be found in almost all of Mitchell's replies to congratulations from other writers, but in the case of Dixon, the neo-Confederate successor to Thomas Nelson Page, she was telling an important truth. Dixon, a North Carolina lawyer turned Baptist minister and itinerant lecturer, had been struck with furious indignation on seeing a stage production of *Uncle Tom's Cabin* in 1901, and within a year had published a refutation of Stowe's charges against his homeland: *The Leopard's Spots,* subtitled *A Romance of the White Man's Burden 1865–1900,* and dedicated to Dixon's wife "Harriet, Sweet-Voiced Daughter of the Old-Fashioned South." Dixon reemployed several "Uncle Tom" characters in a story that involved a hideous rape ("Scarcely a day passed in the South without the record of such an atrocity," the author informs us) and a consequent lynching—the rape unseen but the lynching, actually more a burning at the stake, described in vivid detail. The book reached its climax in the hero's speech to the North Carolina Democratic Convention: "*Resolved,* that the hour has now come in our history to eliminate the Negro from our life and reestablish for all time the government of our fathers."

The Clansman, Dixon's next novel, appeared in 1905. The second volume in what the author called his Race Conflict trilogy, it was published as a tribute to the heroic Ku Klux fraternity of the South's mythic past—a fraternity that had been disbanded decades before. Here Dixon remixed his brew of honey and poison into a story of how "civilization" was saved from Reconstruction's attempt "to Africanize ten great States of the American Union" only through the heroic efforts of "the reincarnated souls of the Clansmen of Old Scotland." Emboldened by his earlier success, Dixon now brought the obligatory rape—of the secondary, expendable heroine—out of the bushes and onto the page, where "the black claws of the beast sank into the soft white throat and she was still." The girl, in shame, chooses to leap off a cliff, hand in hand with her mother. The Clansmen's revenge is swift.

In a touring stage version of *The Clansman* that followed immediately upon the novel's success—Dixon had learned to fight "Uncle Tom" with its own weapons—real horses bearing the girl's avengers charged across the stage as counterparts of Stowe's famous snapping bloodhounds (which in fact were not Stowe's but an invention of the theatre). In Atlanta, Dixon himself came onstage at the end of the show to lecture a responsive audience on his work's historical merit. That same year, in Georgia, separate public parks for whites and blacks were designated, in what was still a novel gesture for a state legislature.

The next summer, again in Georgia, a fierce gubernatorial primary campaign fought on a platform of black disenfranchisement was followed by an outbreak in Atlanta newspapers of an "epidemic of rape," a series of stories played out in rabid headlines and special editions (the winning candidate was also editor of the *Journal*). These newspaper reports led to a five-day wave of white riots, during which mobs of avengers estimated to number ten thousand "killed or tried to kill every negro they saw."

"It will not do to express opinions too freely about the action of the mob in falling on inoffending negroes, for every man you meet justifies it and is enraged": these words, and those just above, are from a letter written by Margaret Mitchell's father—one of a series of letters in which Eugene Mitchell informed his wife, then visiting in New York, about the events of that terrible week in September 1906. (They are quoted in an article published only in the late 1980s by Joel Williamson.) Mitchell weathered the major upheavals by remaining locked in his house with his children—Margaret was nearly six years old—all of them terrified by "a thousand rumors" that "negro mobs had been poised to burn the town" and "cut the water pipes," and fretting because they had no gun. He reported how "Margaret suggested that Mr. Daley's sword," apparently a family relic, "would be a good thing." At the start of the violence, Mitchell had tallied Atlanta's immediate losses, recording that "sixteen negroes had been killed and a multitude had been injured," and at its end he surveyed some of the broader social effects: "Negroes are taking off their hats who never knew they had hats before."

Among the many stagings of the final book in Thomas Dixon's trilogy—*The Traitor,* subtitled *A Story of the Fall of the Invisible Empire*—was one that took place just a few years later in the Mitchell sitting room, with eleven-year-old Margaret as producer and director. Her report of the event to Dixon some twenty-five years afterward details with chilling adorableness how "the clansmen were recruited from the small-fry of the neighborhood, their ages ranging from five to eight," how they wore their fathers' shirts with the "tails bobbed off," and how she herself had to take a male role because none of the little boys would play a part "where they had to 'kiss any little ol' girl.' "

The children's dress-up games of Southern history became big and far more chilling games, for grown-ups, when, in 1915, Dixon's *The Clansman* was adapted and transformed by D. W. Griffith into *The Birth of a Nation,* the first masterpiece of American film. Griffith made something mystic and drivingly apocalyptic out of Dixon's kitchen-garden racism, and the film had a traumatizing impact everywhere in the country—though nowhere, perhaps, so much as in Atlanta. There, in yet another example of the strange interplay of history and fiction which has formed the South, the anticipation of the film's arrival inspired a band of Klan legend-keepers to climb nearby Stone Mountain and, in an elaborate ceremony, set fire to a large cross they placed on its summit. The burning cross was not, in astonishing fact, among the trappings of the original Klan but a poetic addition of Dixon's ("Issue your orders and despatch your courier to-night with the old Scottish rite of the Fiery Cross"), and it was derived—predictably, perhaps, and yet most shockingly of all the South's pseudohistorical traditions—from Walter Scott's *The Lady of the Lake.* In 1915 this fictitious symbol was used to mark the founding of a new Klan, based in Atlanta, which would live longer and cut deeper into the flesh of the nation than its predecessor had ever done. In the local papers, ads for the new organization and the new movie ran side by side.

The Birth of a Nation was surely Margaret Mitchell's model in epic form. (A bulletin from one of her informants in Hollywood boasted that Selznick's street scenes were so fine as to approach even Griffith's.) Her incendiary vision of Reconstruction demonstrates the film's direct visual imprint, as in her confident description of "these negroes" elected to the state legislature, "where they spent most of their time eating goobers and easing their unaccustomed feet into and out of new shoes." In the public mind such descriptions persisted, unquestioned as history, well after the work of scholars like John Hope Franklin and C. Vann Woodward, in the forties and fifties, revealed them as the distortion and propaganda of the Jim Crow years. Knowledge and common knowledge are two different things, and no historian ever had anything like the audiences of Dixon or Griffith or Mitchell. Always proud of the range of study that went into her work, Mitchell was the kind of writer who was able to provide three historical references from her notes for the use of a toothbrush in 1868. But she also attested: "As I had not written anything about the Klan which is not common knowledge to every Southerner, I had done no research upon it."

Some claim has been made for Mitchell as socially progressive, in her context, because her racial portraits and politics are generally devoid of the lust and the terror that animated those of Dixon and Griffith. (Indeed, it isn't certain that the aborted crime at Shantytown was to be a rape, rather than merely a robbery.) Hers is the mildness of complacency, of a work written in and for a time when the dirtiest job had been done—when the Southern situation had been so nearly returned to antebellum conditions that the Klan, with a national membership of more than four million by the early twenties, had to turn its attention to the dangers presented by Catholics and Jews. Yet still it was necessary that the former slaves, if they were no longer to be portrayed as dangerous brutes, be seen as childish clowns in need of protection: how else maintain the glory of the Old

Ways? The heroic Big Sam, who rescues Scarlett from the clutches of Shantytown, is made ridiculous through Mitchell's physical descriptions, all rolling eyes and flashing teeth and "watermelon-pink tongue." This is the bargain Mitchell had to strike in order to give up the virulence of Dixon (whose Sam would have been the rapist) and yet keep her racial and historical righteousness intact.

Charles Everett, Macmillan's outside reader, had added to his useless protest against the Klan episode the suggestion that "the author should keep out her own feelings in one or two places where she talks about negro rule" and the delicate observation that "to refer to Mammy's 'ape face' and her 'black paws' seems unnecessary." Mitchell, ever the lady, thanked her publisher for calling her attention to these matters—"I have tried to keep out venom, bias, bitterness as much as possible"—and, having "meant no disrespect," vowed to make the necessary changes. Clearest among these is the simian substitution in the description of Mammy that follows Ellen O'Hara's death, her "kind black face sad with the uncomprehending sadness of a monkey's face."

Bias and bitterness, though, characterize Mitchell's entire account of Reconstruction. To question this attitude is not to deny the privation, the ruin, the real suffering of Southern gentry—Edmund Wilson compared the crushing of the South during and after the war to the crushing of Hungary in 1956—but, rather, to underline Mitchell's inability to see that time and place from any but one point of view, or to admit the complexity of the situation or of the truth. W. E. B. Du Bois, writing in 1935, set forth the matter simply: "One fact and one alone explains the attitude of most recent writers toward Reconstruction; they cannot conceive of Negroes as men."

Big Sam, when he encounters Scarlett, is wearing a Union jacket and has just returned from the North, where "dem Yankee folks, fust time dey meet me, dey call me 'Mist' O' Hara," and where he resented and vehemently rebuffed all the questions " 'bout de blood houn's dat chase me an' de beatin's Ah got." All he wants, he says, is to go back to Tara the way it was: "Ah done had nuff freedom." There is a literary tradition for Sam's homesickness—even a black one, as in Paul Laurence Dunbar's "You kin jes' tell Mistah Lincum fo' to tek his freedom back"—but it does not seem to be supported by the reality. Among the thousands of aged former slaves who were interviewed in the WPA Federal Writers' Project of the thirties and early forties, when the South was the nation's foremost economic problem and their living conditions were generally abysmal, the question of Emancipation was answered as in a single voice. In the plain statement of a man who went by the name of Moses Mitchell, "Here's the idea: freedom is worth it all."

Despite the issues it raises, Margaret Mitchell's all-American bestseller is not fundamentally concerned with politics, or, for all its exertions, with race, except insofar as these subjects set its scenes and affect its characters. It is unimaginable that the pleasure so many readers have found in the book bears any profound relation to these public subjects. What *Gone with the Wind* is ultimately about is romance and sex—these subjects, rather than a female point of view, are what made it a "woman's book"—and there is no surer demonstration of this fact than the false alarm of the scene outside Shantytown. Rape is not a matter of politics in Mitchell's reconstructed South. When it occurs, in one of the most carefully prepared and climactic scenes of the book, the rapist is not black but white, not a monster but a handsome hero, and the heroine is neither murdered nor avenged but awakened to pleasure and a taste of victory.

Rhett Butler is a pure projection of idealized male sexuality, an ever-potent cliché. Although the author at no time presumes to enter his head—leaving him free of doubt, error, foolishness—his physical properties are subject to endless description and evaluation. He first appears at the big eve-of-war barbecue, tall, older, and "powerfully built": "Scarlett thought she had never seen a man with such wide shoulders, so heavy with muscles, almost too heavy for gentility." There is "a look of good blood in his dark face," and both the quality of the blood and, especially, the darkness are frequently rediscovered and remarked.

"Swarthy as a pirate," his "animal-white teeth" flashing, Rhett Butler is clearly a descendant of Don Juan and of Heathcliff, but above all, he is the son of the Sheik: Margaret Mitchell had seen Rudolph Valentino's first starring film, *The Four Horsemen of the Apocalypse,* often enough to derive her scandalous charity-ball Apache dance from it, and in 1924 she had interviewed the Latin actor for the *Atlanta Journal.* "He seemed older—just a bit tired," Mitchell informed her Sunday-supplement readers. "His face was swarthy, so brown that his white teeth flashed in startling contrast to his skin, his eyes tired, bored but courteous." Valentino had by then become, via *The Sheik,* the biggest male sex star of the movies, having deposed at a stroke all the good-Joe heartthrobs already on the scene by means of a newly dangerous, predatory—not to say swarthy—glamour. Some have speculated that Valentino came as close as acceptable standards permitted to suggesting the forbidden erotic appeal of a black man; the speculation alone makes Rhett Butler seem a more complex psychological creation than Mitchell's prose ever allows, and may make us read her hymns to his virility with a warier eye. Mitchell's Rhett is a "pagan prince" whose movements suggest a "pagan freedom and leashed power"; there is about him something "almost sinister," a "suave brutality." His body is dangerous, even when elegantly dressed, and its violence is ever implicit, as when he joins Scarlett for a carriage ride: "The muscles of his big body rippled against his well-tailored clothes, as he got in beside her, and, as always, the sense of his great physical power struck her like a blow."

By the time Scarlett O'Hara becomes Mrs. Butler, she has already been married twice, and all the while has chastely adored the golden, honorable, and married Ashley Wilkes. But the pirate (grandson of a pirate, actually) has aroused something in her, something incomprehensible to her, beginning with an insulting glance that makes her feel "that her dress was too low in the bosom" and annoys her most "because she did not feel insulted"; continuing with the first seductive move—his lips to her palm—which brings up a "treacherous warm tide of feeling"; and going on through the elaborate spirals of a woman's sexual choreography to the poster moment where, above the flames of Atlanta, "he bent her body backward and his lips traveled down her throat to where the cameo fastened her basque."

The "Road to Tara," that thematically surging episode in which Scarlett is abandoned by Rhett to lead her helpless charges on alone toward home, is emphatically counterbalanced in the book by an equally famous later scene in which she is swept into Rhett's arms and carried, in fear and protest, up a grand staircase to their bed. The "Road" scene marks our heroine's discovery of her strength and the making of her resolve; the "Rape" (as Selznick, for one, frankly termed it) is her glad surrender of strength to a force greater than her own—"to arms that were too strong, lips too bruising, fate that moved too fast."

Many women who are passionate fans of *Gone with the Wind*—and of both these scenes—have claimed that the word "rape" is not an adequate description of the events that night in the Butler bedroom. The preferred term, not only in this instance but in the thousands of such scenes that fill women's popular fiction, is something along the lines of "forceful persuasion." This is not Sade and it is not Faulkner—not painful, not punishing. Its roots are as old and noble as Richardson's *Clarissa*—the modern novel born in an excruciatingly prolonged contemplation of rape. Margaret Mitchell herself had a variety of models to choose from in the years when she was growing into her vocation; in 1920 Mencken noted that among the few types of books that almost never lost money in the United States were "novels in which the heroine is forcibly overcome by the hero"—a category that placed second only to murder stories.

But the "baby-faced li'l vamp" didn't require fiction to introduce her to the attractions of forceful persuasion. The giddy "It girl" letters that Mitchell sent North after leaving college, collected by Jane Bonner Peacock under the title *A Dynamo Going to Waste,* retail what she called her " 'cutie' career" and brim with the thrills of sexual brinksmanship: "Promised to marry a youthful cave man—just to see what would happen (I found out quite speedily and had a helluva time getting him off the scene for keeps)"; and "I used to have an elegant time in my early youth . . . by giving a life-like imitation of a modern young woman whose blistering passions were only held in check by an iron control. . . . Thoughts of seduction were tabled and rape became more to the point." She even encounters or creates the occasional Rhett: "Ever know a man who makes you acutely conscious that your dress is too low? That's A.B. I suddenly began to loathe him. I took sidelong glances at him, noting his sensual mouth and closely cropped moustache and meeting his assured, faintly sneering eyes." The specimen escorts her home and won't leave, "and then," she wails, "the fun began! . . . When you've liked and trusted a man, it is no pleasant sight to see him lose his head and go wild. It was the evening dress, I guess, and the fact that both straps slipped down at this inopportune time."

It wasn't the mysterious A.B. whom Mitchell married soon after this but another in the same line. Anne Edwards, in her biography of the author, *Road to Tara,* identifies Mitchell's first husband, Red Upshaw, with Rhett Butler, pointing out that even the unexplained middle initial "K" that Scarlett espies on Rhett's handkerchief belongs properly to Berrien Kinnard Upshaw, and could be seen as a kind of personal signal. At the time she was writing her book, Mitchell still slept with a pistol by her bed in case the signal ever again took living shape in her vicinity. Their last meeting is memorialized in nonfictional form in Mitchell's sworn deposition in the case of the Upshaws' divorce proceedings. Married in September 1922, the bride and groom separated within months, and the next July he suddenly reappeared at her home. "Mr. Upshaw demanded his connubial rights," Mitchell testified, "after striking me with his fist upon my left arm about the elbow." Her counsel added that he had "jerked her against a bed, causing her to be bruised all over her body." The maid had come running, and as Upshaw left he had delivered a final punch to the eye; Mitchell was hospitalized for two weeks.

Nothing of the kind happens to Scarlett O'Hara, of course. Fiction is different. And romantic fiction is not only not reality but very nearly its antidote. In Mitchell's latter-day fairy tale, the darkly beautiful pagan prince, drunk and angry, is desperately in love with the delicately beautiful woman who has wronged him cruelly through years of marriage and now stands trembling before him; his cravat and shirt are open, her wrapper is pulled tightly closed. She runs from him but loses a slipper in her flight, and he is suddenly beside her: "He swung her off her feet into his arms and started up the stairs. Her head was crushed against his chest and she heard the hard hammering of his heart. . . . Up the stairs, he went in the utter darkness, up, up, and she was wild with fear." It is a long way up, and a long paragraph. Arriving at last at the landing, he "bent over her and kissed her with a savagery and a completeness that wiped out everything from her mind but the dark into which she was sinking and the lips on hers." The writing is Mitchell at her fanciest: "She was darkness and he was darkness and there had never been anything before this time." And then—they are still on the stairs—"she had a wild thrill such as she had never known. . . . For the first time in her life she had met someone, something stronger than she, someone she could neither bully nor break, someone who was bullying and breaking her."

This neo-Victorian ravishment concludes in accord with the more historically appropriate wisdom of Anita Loos: "In those days," the sibyl of the twenties proclaimed, "a girl could wake up smiling." Scarlett O'Hara wakes up—it's the next thing we know—blushing, and filled with "the ecstasy of surrender." (In the movie, she warbles a little morning-after song; Mitchell's friend Susan Myrick, responsible for period authenticity, had first suggested "It ain't what you do, it's the way that you do it.") She worries about whether she can ever again even imagine herself to be a lady, but most of all she feels newly secure in love. What the "wild, mad night" has meant to her, finally, and what the old-fashioned "forceful persuasion" scene seems to mean generally to heroines and to readers, is proof resplendent of her own desirability. This is her satisfaction, and this is, ironically—"Now she knew the weakness of his armor"—her power.

When *The Clansman* was reprinted in the early 1940s, on the coattails of *Gone with the Wind,* the book's new front cover identified it as a "world-famous love story"; the back cover advertised a line of "Love Stories" in cheap editions, with titles that ranged from *Pride and Prejudice* and *Wuthering Heights* to *Grand Hotel, Prodigal Nurse,* and *Impatient Virgin.* Quite a slope. By the 1950s, the genre of romance fiction—settled into a kind of pinkies-up semipornography—was outselling even mysteries, and by the 1970s it had become an assembly-line product far outselling all other categories of paperbacks, whether because of a reaction against feminism or the ingenuities of marketing or expanding leisure and cultural vacuity it is hard to say.

From national best-sellers to unnameable checkout-counter Harlequins, the "romantic" pattern was largely fixed. An academic study of the proliferating species—a study in the "ethnographies of reading"—by a Duke professor, Janice A. Radway, provides a precisely charted analysis of thirteen standard narrative events, or "functions," that make up the fully evolved contemporary romance novel. Using Kathleen Woodiwiss's particularly popular *The Flame and the Flower* as her main demonstration model—hero Brandon, heroine Heather—Radway outlines the rules of the game: function 1, the heroine's social identity is destroyed; function 2, the heroine reacts antagonistically to an aristocratic male; and so on. In this way, Radway arrives at the ideal romance's midpoint: "Although he continues to believe Heather is an opportunist and she herself remains angry with him over the rape (function 7), their emotional separation does not stop him from surprising her with especially thoughtful gifts (function 8)." No wonder Scarlett assumed she had something to smile about.

But the big upset in Margaret Mitchell's story is her break with the very formula that she both exploited and exemplified: Scarlett O'Hara's brush with sensual bliss is not the start of a long road disappearing into marital contentment. Probably the most rigorous aspect of *Gone with the Wind* is Mitchell's unyielding detachment from her heroine. The author never falls for her creature's charms, and she neither softens her faults nor forgives them. Scarlett's halting moral and emotional development occurs in barely measurable increments: on page 947, she experiences "one of the few adult emotions" she has ever had, and it is not until page 1031, almost the last in the book, that her husband's unwonted show of grief over the death of their child elevates her to the point where for "the first time in her life she had ever been sorry for anyone without feeling contemptuous as well, because it was the first time she had ever approached understanding any other human being." It is, of course, too late.

Rhett Butler's departure and his line "My dear, I don't give a damn" have vexed many a sentimental heart. (The rhythmic upbeat of "Frankly" was added to the movie, and Selznick, famously, had to fight for the right to "damn.") Mitchell freely told everyone that she had written the ending first; she knew exactly what she was after. Considering the contrast between the author's reckless first marriage and the resigned respectability of her second, and looking at the photographs in which she hardens and dries so conspicuously through her twenties and her thirties—as though Clara Bow were turning into Norma Shearer—one might imagine the revenge of a premature dowager on her own all too audacious youth. Yet even in the middle of that youth Mitchell seems to have had such an idea, such an ending, in mind. In letters written just after her time at Smith, she complains of her inability to complete a story she is working on, which is to conclude with a kiss at a wedding—a kiss by which a man deliberately lets a woman know that by marrying someone else she has just ruined her life. The fledgling author was having no end of trouble describing that all-meaning kiss, but she did know that "when the insistent demand of his lips on hers makes her admit that she always would be his, then he'd leave." (To which she appended, "I do see vast possibilities for 'hot stuff' in that passage!")

When Macmillan's reader suggested that Mitchell reduce the sense of finality in Rhett's departure, she replied, "I'll change it any way you want, except to make a happy ending." Her intention, she firmly stated, was "to leave the ending open to the reader." And so it has come to pass that Americans for more than half a century have divided—as other nations divide into Platonists and Aristotelians, into Tolstoyans and Dostoevskyans—into those who fully expect Rhett Butler to return to Scarlett O'Hara and those who know for certain that he never will.

On the seventy-third anniversary of the Battle of Atlanta, in July 1937, a Confederate flag was flown over the old city center at Five Points for the first time in general memory; it was the beginning of a new—or renewed—tradition. In a letter written that November, Mitchell reported to her publisher that "the book is on the required collateral reading lists of many high schools in the South, and even a number of junior high schools and grammar schools are using it." She noted that "the sex angle" seemed to pass easily over the youngest heads, or had proved amenable to such interpretations as that of one "bright child"—aged eleven—whose school book report referred to Belle Watling's brothel, in the scene where the heroic Klan members are sheltered there, as "the swankiest night club in Atlanta." This happy example, she wrote, had given her a sense of relief: "After that I had no fears that I was polluting the youthful mind."

What kind of moral responsibility does she bear, after all? The United States is a young country, and many have accused her of polluting the *national* mind—or, at least, of stirring up and adding cheap perfume to what has always lurked at the bottom of it. Accusations from the Northern and leftist press that her book amounted to "negro baiting" drew her most indignant, fluttering responses: "Personally I do not know where they get such an idea. . . . The negroes in this section have read it in large herds and while I have not heard as many comments as I would like to hear, my friends are continually telling me what colored elevator operators, garage attendants, etc., tell them and these colored people seem well pleased."

What seems clear is that Margaret Mitchell represented her time and place as accurately as she had intended Scarlett O'Hara to do: she was outraged innocence, she was the most staunchly and unquestioningly segregated society since slavery, she was her own Atlanta. In 1926, the year she began writing *Gone with the Wind,* a new city ordinance prohibited Negro barbers from serving white women or children; in 1940, the year after the movie's release, another ordinance divided the city's taxis according to the permitted race of passengers. And in the years between, she wrote a love story that made such practices seem as natural and as familial as weddings and funerals.

It has been said that while the North won the war, Margaret Mitchell won the peace, and certainly Appomattox was no match for the big guns of Hollywood. The movie was carefully designed to soften many of the book's direct racial blows, except in the case of the dizzyingly imbecilic Prissy. ("I did everything they asked me to," Butterfly McQueen said, "except I wouldn't let them slap me and I wouldn't eat the watermelon . . . I hated the part then.") David Selznick expanded Mammy's role (McQueen was counterbalanced by the august Hattie McDaniel) and insisted that, in sum, "the Negroes come out decidedly on the right side of the ledger." But the seductiveness of the film swept all before it: by force of color and music, by the dynamics of movement and incident, and by the beauty of movie stars, *Gone With the Wind*—Selznick and Vivien Leigh's *Gone With the Wind*—became a glorification of the Old South such as had never been seen before.

On the eve of the Atlanta premiere, in December 1939, a celebratory Junior League costume ball for six thousand managed to recapture, in the words of a local paper, "the days at Tara Hall, when every man was a master and every man had a slave." The entertainment that night alternated between the "hot music" of Kay Kyser's swing band and the spirituals of the Reverend Martin Luther King, Sr.'s Ebenezer Baptist Church choir, which performed in slave attire before a plantation backdrop and which included among the onstage "pickaninnies" the minister's ten-year-old son, Martin Jr. Mitchell did not attend the ball, but she was present, in full glory, at the next night's opening. And give or take a few architectural details, the world of the movie was just as she had envisioned it: an Eden that knew no serpent until the Yankees came.

Selznick had from the start ruled out any presentation of the Klan, voicing fear that its appearance "might come out as an unintentional advertisement for intolerant societies in these fascist-ridden times." Even before his cameras were rolling, the book was on its way to becoming one of Germany's biggest sellers. At the time of the American Civil War, the German government, in the person of Bismarck, had identified with the triumphant North and the struggle to remain unified, but after 1918 Germany was no longer identifying with victors. One applicant for the job of translating *Gone with the Wind* into German stated in a letter to Macmillan that "it does not contain any ideas which could displease the Hitler government" (unlike the once popular works of Upton Sinclair, John Dos Passos, and Sinclair Lewis, which had been thrown onto the bonfires of 1933). *Vom Winde Verweht* had sold more than 360,000 copies by 1941, at which time its message was revealed to the Nazi government as so mercurial, its value as propaganda so unreliable, that it was suddenly banned. The Germans, after allowing the book to appear in occupied countries, had discovered that they were not the only ones to identify with the rebels in gray who would not accept defeat. Mitchell received reports that her book was serving as "a great morale builder" in those countries, and that, as she wrote to her publisher, "occupied nations identified themselves with the South during Reconstruction, identified the Ku Klux Klan with the forces of the Resistance, and were heartened by the thought that the South eventually got back its own state governments." In 1944 the *New York Journal-American* reported that bootlegged copies of the book were selling for sixty dollars in France and for nearly as much in Holland, Norway, and Belgium. According to the report, orders to seize all such copies had gone out, and people caught with the book in their possession were being shot.

▪ ▪ ▪

A more current (and more credible) proof of the force of Mitchell's creation was its ability to keep the official sequel, Alexandra Ripley's *Scarlett,* on the best-seller list for the better part of a year after it was published, in 1991; indeed, this success indicates a near savage interest in finding out, after fifty-five years, what happened. Sequels or plans for sequels had abounded since 1936, and Mitchell was forever squelching "last chapter" contests, as well as whole manuscripts with titles like "Return of the Gentle Wind" or "Whispering Winds." She stated flatly and often that she would never attempt a sequel herself, and her husband stood constant legal guard over the issue. "Not only would such a sequel be an 'unfair appropriation' of her skill," he wrote to Macmillan in 1938, "but it would also damage her through being an 'inferior imitation' (if we may judge by the sequels already written)."

Perhaps they should have just let it happen. What was finally made of the job seems almost inevitable, given the course of the woman's romance novel—Scarlett not as belle or vamp but as Cosmo girl, and suffering from painfully hardened prose implants. "I mustn't ever tell him again that I love him," she reproves herself. "That makes him feel pressured." This Scarlett hurries along thick carpets to embroidered bellpulls; she orders ice swans for her parties, and cases of champagne ("Scarlett did so like for things to be stylish"); she is a courageous shopper. But, writing aside, and critics aside, it is hard to believe that this "inferior imitation" offered satisfaction to many readers, if only because no real connection was ever made with Mitchell's characters. Call them Rhett and Scarlett, call them Hamlet and Ophelia, these newcomers are not for a moment any other pair but Brandon and Heather. Of Mitchell's world the only recognizable sign is in the infamous injunction of her estate (a consortium of nephews and lawyers) against the presentation of explicit sex scenes, homosexuality, or miscegenation. F. Scott Fitzgerald said of *Gone with the Wind,* "I felt no contempt for it but only a certain pity for those who considered it the supreme achievement of the human mind." In present company it may seem to its worshippers more supreme than ever.

A strenuous effort to throw off the stigma of decades of literary patronizing was made in a 1991 biography of Mitchell, Darden Asbury Pyron's *Southern Daughter.* As a historian, a Southerner by breeding and conviction, and a true *Gone with the Wind* believer, Pyron was well qualified for the task, but his attributes don't always cohere, and at its extremes, his book veers between a bright-eyed folksiness—"Success slammed through the

Mitchell-Marsh apartment on Seventeenth Street like Huns and Tartars," he tells us—and a tendentious academic voice, of which no sample is needed. He emerged with a thorough and exacting but relentlessly earnest study that falters through the lack of any major new material to present—Mitchell's heirs destroyed everything within their reach—and a consequent falling back, for substance and novelty, on overstated theories about the psychological subtext of the novel.

Pyron's book almost seems a counterbiography to the popular Anne Edwards 1983 account—popular both in its sales and in its tradition. Edwards, the author of her own unpublished sequel (it was part of a film project that failed) and also of biographies of Vivien Leigh and Judy Garland, wrote a clear and straightforward (dare one say "readable"?) story. Though it was generally well researched, and was the source of a number of the narrative incidents reported in the foregoing, Edwards's work suffered from errors based on misreadings of the newly available Macmillan Archive (she mistakenly suggests that Mitchell envisioned a happy ending) and an occasional stumble from a usually sensible tone into the likes of "Fury rose inside Peggy Mitchell's small-breasted chest." Still, her book seemed, overall, well suited to its subject.

Pyron's far weightier work suggests—almost seems to require—a darker figure at its center. He makes a great deal of a succession of illnesses and accidents that plagued Mitchell's life (her letters offer a barrage of boils and broken bones, of collisions with cars and with her furniture), and he takes her somewhat loose claim that she began writing her novel only because "I couldn't walk for a couple of years" as proof that, in his words, "she never failed to associate her fiction with disease and suffering." Mitchell's experience of physical pain was, according to Pyron, a determining element of what she finally wrote, and the writing itself thus becomes a story fulfilled only in "a chronicle of horrors—death, abandonment, rejection, alienation, smashed hopes, and fatal misunderstanding." Convinced that his heroine has been done wrong by a Northern literary establishment that despised her politics and dismissed her art, Pyron is nothing if not zealous in supplying the required interpretative shadows—the "existential pessimism"—to prove to these critics that there was something to dismiss.

More persuasive shadows are brushed in with Pyron's portrait of Mitchell as an avid collector of pornography, a matter that was no great secret at the time among her friends; in a letter to her sister-in-law she refers easily to "a dog-eared copy of *Elsie Dinsmore* which, for appearances,

I keep shelved between copies of *Jurgen* and *How Kate Lost Her Maiden Head* (a most informative volume)." Mitchell particularly doted on the case studies of Havelock Ellis, which she ordered from the "dirty book stores" in New York. A friend with whom she shared her treasures—at one time, she belonged to a kind of Atlanta hobby club—reported that her favorite Ellis history involved a case of male lust so extreme as to enforce disregard for its object's sex or species. As Pyron puts it, rather succinctly, "she dreamed of satyrs."

The point of Pyron's discussion is not to soil Mitchell's reputation but, on the contrary, to add dimension to it—to reveal her as a complex twentieth-century figure. (Perhaps he had in mind R. W. B. Lewis's revelations about Edith Wharton.) But Pyron stops at any exploration, or even acknowledgment, of the series of masks that Mitchell seems to have worn, of the elaborate playacting that characterizes her life—a path that would lead back, inevitably, to the treacherous issue of race. He bridles at the suggestion that an early story of hers turned upon miscegenation. (Apparently an attempt at Faulknerism, the story was destroyed after her death; the heroine, it might be noted, was named 'Ropa—short for Europa, of mythical godly ravishment.) It was Pyron who uncovered the episode of Mitchell's quitting the history course at Smith, but he minimizes its import with an observation on "her discomfort about living in the North."

In the end, though, and most curiously, Pyron crushes all the various modest possibilities of interpretation of Mitchell's book under a single monumental theory. After revealing the depths of the author's sexual preoccupations, the biographer proposes a primary interpretation of *Gone with the Wind* as a search for the Mother—as a disquisition on the difficult relationship of mother and daughter. In this view, Rhett not only is not Mitchell's first husband (Pyron briskly dismisses Edwards's identification) but isn't even a man; rather, he is a stand-in for Maybelle Mitchell. And so, for good measure, is Tara: "the mythic mother . . . Maybelle Mitchell." Indeed, Pyron's *Gone with the Wind* is a romance not between a hero and a heroine but between a woman and her mother, and the meaning of the novel is to be found, he reports, in Mitchell's sense of "the central, defining characteristic of women's lives—the birth experience." Aside from the reductionism of this notion, which seems almost purposefully at odds with the lessons that Maybelle Mitchell tried to teach her daughter, one can only wonder how it is to be reconciled with the life of the childless author itself.

If neither Edwards nor Pyron quite succeeds in making Mitchell a three-dimensional figure, it must be acknowledged that they had little to work with, not only because so much material was destroyed but because of the absence in what abundantly remains of any sign that the woman developed her thoughts, that she reflected or reconsidered, or, indeed, that she had any interior life. This may be a result of deliberate secrecy and obfuscation, or perhaps there is some connection, after all, with the embrace of the accepted and the clichéd which marks her writing. Over the years, Mitchell came to seem increasingly mechanical in her responses, fierce about copyrights, and terrified of being forgotten. "I think the war, of course, had something to do with the cessation of public interest in me," she wrote in the fall of 1940, "and the election naturally diverted attention." Her death in 1949—she was run over by an off-duty taxi driver—was front-page news. Just a few months earlier, she had written a collegial note to William Faulkner—her first—saying that she thought he might like to see a reproduction of the Italian jacket cover of *Sanctuary* in a catalogue she had come across: "I showed it to a friend who is a great admirer of your books—'Dear me—how *explicit* the Italians are!' " There is no record that he replied.

Margaret Mitchell is unlikely ever to join Faulkner or, for that matter, Harriet Beecher Stowe in the gleaming uniform rows of the classic Library of America editions. In the history of American literature—in all the published histories—her place, when she has one, is in a corner apart, as a vulgar aside having to do with numbers rather than words. She doesn't even make it onto the list of Best Civil War Novels in either of the studies devoted exclusively to the genre. And except among the sociologists of best-sellers, she has been as fully excluded from current reconsiderations of women's writing. *Gone with the Wind* hasn't a place in anyone's canon; it remains a book that nobody wants except its readers.

More than six decades after its publication, Mitchell's novel is an accepted American artifact, and still a symbol of the gaping cultural divide that it once helped to define. As the great unnamed, *Gone with the Wind* hovers over a richly revealing 1962 essay on our national myths by Malcolm Cowley, Mitchell's longtime and most perceptive adversary. Cowley lists a succession of post-1920 literary legends: T. S. Eliot's spiritual wasteland, Fitzgerald's Jazz Age, Hemingway's Lost Generation, Erskine Caldwell's Tobacco Road, Steinbeck's Okies, and the "Southern cavalier legend," which, although more than a century old, "was raised to a new dimension by William Faulkner." Needless to say, the red earth of Tara—even from this Olympian perspective—is not within Cowley's view, although he does go on to state that "it was during this period, too, that Troy was burned again in the shape of Atlanta." But in what book? Written by whom? The eminent critic concludes, "Hundreds of authors working in collaboration had given us another Iliad, of sorts." An American epic, then—of sorts. An American embarrassment, reflecting a society, an era, a nation: our Dunciad, our Scarlettiad. Blatant, commercial, disowned. There is, after all, some spark of justice in the fate of Margaret Mitchell's blundering colossus, condemned by posterity to live on triumphantly yet always separate and never, never equal.

A Society of One

Zora Neale Hurston

In the spring of 1938, Zora Neale Hurston informed readers of the *Saturday Review of Literature* that Mr. Richard Wright's first published book, *Uncle Tom's Children,* was made up of four novellas set in a Dismal Swamp of race hatred, in which not a single act of understanding or sympathy occurred, and in which the white man was usually shot dead. "There is lavish killing here," she wrote, "perhaps enough to satisfy all male black readers." Hurston, who had swept onto the Harlem scene a decade before, was one of the very few black women in a position to write for the staunchly traditional *Saturday Review.* Wright, the troubling newcomer, had already challenged her authority to speak for their race, and had set the terms on which they battled. Reviewing Hurston's novel *Their Eyes Were Watching God* in *The New Masses* the previous fall, he had dismissed her prose for its "facile sensuality"—a problem in the history of "Negro expression" that he traced to the first black American female to earn literary fame, the slave Phillis Wheatley. Worse, he accused Hurston of cynically perpetuating a minstrel tradition meant to make white audiences laugh. It says something about the social complexity of the next few years that it was Wright who became a Book-of-the-Month Club favorite, while Hurston's work went out of print and she nearly starved. For the first time in America, a substantial white audience preferred to be shot at.

Black anger had come out of hiding, out of the ruins of the Harlem Renaissance and its splendid illusions of societal justice willingly offered up to art. That famed outpouring of novels and poems and plays of the twenties, anxiously demonstrating the Negro's humanity and cultural citizenship, had counted for nothing against the bludgeoning facts of the Depression, the Scottsboro trials, and the first-ever riot in Harlem itself, in 1935. The advent of Richard Wright was a political event as much as a literary one. In American fiction, after all, there was nothing new in the image of the black man as an inarticulate savage for whom rape and murder were a nearly inevitable means of expression. Southern literature was filled with Negro portraits not so different from that of Bigger Thomas, the hero of Wright's 1940 bombshell, *Native Son.* In the making of a revolution, all that had shifted was the author's color and the blame.

As for Hurston, the most brazenly impious of the Harlem literary avant-garde—she called them "the niggerati"—she had never fit happily within any political group. And she still doesn't. In this respect, she was the unlikeliest possible candidate for canonization by the black and women's studies departments of our own era. Nevertheless, since Alice Walker's "In Search of Zora Neale Hurston" appeared in *Ms.* in 1975, interest in this largely forgotten ancestress has developed a wild momentum. All her major work has been republished (most recently by the Library of America), she is the subject of conferences and doctoral dissertations, and the movie rights to *Their Eyes Were Watching God*—which has sold more than a million copies since 1990—have been bought by Oprah Winfrey and Quincy Jones. Yet despite her new, almost sanctified status, Hurston's social views are likely to seem as obstreperous today as they did sixty years ago. Looking at her difficult life and often contradictory legacy straight on, it is nearly impossible to get this disarming, often deceptive conjure artist to represent any cause except the freedom to write what she wanted.

Hurston was at the height of her powers when, in 1937, she first fell seriously out of step with the times. She had written a love story—"Their Eyes Were Watching God"—and become a counterrevolutionary. Against the tide of racial anger, she wrote about sex and talk and work and music and life's unpoisoned pleasures, suggesting that these things existed even for people of color, even in America; and she was judged superficial—by implication, merely feminine. In Wright's account, her novel contained "no theme, no message, no thought." By depicting a Southern small-town world in which blacks kept mostly to themselves, enjoyed their own rich cultural traditions, and were able to assume responsibility for their own lives, Hurston appeared a blithely reassuring supporter of the status quo.

The "minstrel" charge was aimed most furiously, however, not at what she said but at how she said it. Black dialect was the very substance of Hurston's work, and that was a dangerous business. Disowned by the founders of the Harlem Renaissance for its association with the shambling, watermelon-eating mockeries of American stage convention, dialect remained an irresistible if highly self-conscious resource for black writers from Langston Hughes and Sterling Brown to Wright himself. But the feat of rescuing the dignity of the speakers from decades of humiliation—so recently reinforced by the nation-sweeping success of *Gone with the Wind*—required a rare and potentially treacherous combination of gifts: a delicate ear (Hurston derided Wright's use of dialect as tone-deaf) and a generous sympathy, a hell-bent humor and a determined imperviousness to shame. All this Hurston brought to *Their Eyes Were Watching God*—a book that, despite its slender, private graces, aspires to the force of a national epic, akin to the achievement of Mark Twain or Alessandro Manzoni, offering a people their own spoken language freshly caught on paper and raised to the heights of poetry.

"It's sort of duskin' down dark," observes the otherwise unexceptional Mrs. Sumpkins, checking the sky and issuing the local evening variant of rosy-fingered dawn. "He's uh whirlwind among breezes," one front-porch sage notes of the town's mayor; another adds, "He's gut uh throne in de seat of his pants." The simplest men and women of the all-black town of Eatonville have this wealth of images easy at their lips, while a more thoughtful type assures his jealous girlfriend that "de girl baby ain't born and her mama is dead, dat can git me tuh spend our money on her." This is dialect not as a broken attempt at higher correctness but as an extravagant game of image and sound. It is a record of the unique explosion that occurred when African people with an intensely musical and oral culture came up hard against the King James Bible and the sweet-talking American South, under conditions that denied them any outlet for their visions and gifts except the transformation of the English language into song.

▪ ▪ ▪

Hurston was born on the lands of a former plantation in tiny Notasulga, Alabama, to a family of sharecroppers in 1891—about ten years before any date she ever admitted to. Both her biographer, Robert E. Hemenway, and her admirer Alice Walker, who put up a tombstone in 1973 to mark Hurston's Florida grave (inscribed " 'A Genius of the South' 1901–1960 Novelist, Folklorist, Anthropologist"), got this basic fact as wrong as their honored subject would have wished. Hurston was a woman used to getting away with things: her second marriage license lists her date of birth as 1910. Still, the ruse stemmed not from ordinary feminine vanity but from her desire for an education and her embarrassment at how long it took her to get it. The lie apparently began when she entered high school, in 1917, at twenty-six.

She had been very young when the family moved to Eatonville, Florida, the first incorporated black town in America (by 1914 there would be some thirty of them throughout the South), in search of the jobs and the relief from racism that such a place promised. In many ways, they found precisely what they wanted: John Hurston became a preacher at the Zion Hope Baptist Church and served three terms as mayor. His daughter's depictions of this self-ruled colored Eden have become legend, and in recent years have even seemed to hold out a ruefully tempting alternative to the ordeals of integration. The benefits of the self-segregated life have been attested by the fact that Eatonville produced Hurston herself: a black writer uniquely whole-souled and self-possessed and imbued with (in Alice Walker's phrase) "racial health."

Her mother taught her to read before she started school, and encouraged her, she said, to "jump at de sun." Her father routinely smacked her back down and warned her not to act white; the child he adored was her docile older sister. One must go to Hurston's autobiographical novel, *Jonah's Gourd Vine,* for a portrait of this highly charismatic but morally weak man, whose compulsive philandering eventually destroyed his family and all he'd built. The death of Zora's mother, in 1904, began a period she would later seek to obliterate from the record of her life. In *Jonah's Gourd Vine* the bold younger daughter is summoned to her mother's deathbed for a fearful, parting admonition: she is to get all the education she can, and to make sure that she never loves anyone more than she loves herself. Although Hurston's actual autobiography, *Dust Tracks on a Road,* is infamously evasive and sketchy about what followed (burying a decade does not encourage specificity), it does acknowledge her having been thrown out of a Jacksonville school when her father stopped paying the bills, and then having been shunted among her brothers' families with the lure of more schooling always giving way to cleaning house and minding children. And all the while, she recalls, "I had a way of life inside me and I wanted it with a want that was twisting me."

Working every kind of menial job—maid, waitress, manicurist—she managed to finish high school by June 1918 and went on to take courses at Howard University, where in 1921 she published her first story, in the literary-club magazine. Harlem was just then on the verge of vogue, and the Howard club was headed by Alain Locke, founding prince of the Renaissance, a black aristocrat out of Harvard on the lookout for writers with a sense of the "folk." It was what everybody would soon be looking for. The first date that Hurston offers in the story of her life is January 1925, when she arrived in New York City with no job, no friends, and a dollar and fifty cents in her pocket—a somewhat melodramatic account meant to lower the lights behind her rising glory.

One story had already been accepted by *Opportunity*, the premier magazine of "New Negro" writing. That May, at the first *Opportunity* banquet, she received two awards—one for fiction and one for drama—from such judges as Fannie Hurst, a best-selling, four-handkerchief novelist with a liberal interest in race relations, and Eugene O'Neill. Hurston's flamboyant entrance at a party following the ceremonies, sailing a scarf over her shoulder and crying out the title of her play—"Color Struck!"—made a greater impression than her work would do for many years to come. This was the new, public Zora, all bravado and impudent laughter, happily startling her audience with the truth of its own preoccupations.

That night, she attached herself to Fannie Hurst, for whom she was to work off and on in the next year, beginning as a secretary and then, when it turned out that she couldn't type or keep anything in order, as a kind of rental exotic, complete with a supply of outlandish down-home stories and a turban. (Her new boss once tried to pass her off in a segregated restaurant as an African princess.) Hurston's Harlem circle was loudly scornful of the part she was willing to play. For her, though, it was experience; it was not washing floors, it was going somewhere. And the somewhere still hadn't changed. At the banquet she had also met Annie Nathan Meyer, a founder of Barnard College, from whom she soon finagled a scholarship.

And in the fall of 1925, the ever-masquerading, glamorous, Scott-within-Zelda of Lenox Avenue became a schoolgirl again—Barnard's only black student, apparently hardly older than a freshman—and discovered that her stories might also be considered (and respected) as a part of the still-new field of anthropology.

Hurston dived headlong into this exciting realm of thought, which had been conceived principally by her teacher, Franz Boas, a German-Jewish immigrant who'd founded the department at Columbia in the first years of the century. (Like all his students, Hurston called him Papa Franz, and he teased that of course she was his daughter, "just one of my missteps.") Boas, almost alone in an age of rampant "scientific" racist theorizing in Europe and the United States, based his work on a bedrock belief in the adaptability and variability of so-called essential racial characteristics. In his frankly political, profoundly humanist (if also carefully "scientific") studies, he insisted that such factors as culture and learning and even diet have as much influence on human development as heredity, and he set out to prove how close the members of the family of man might really be. He had already examined the changing characteristics of the children of European immigrants—what might be called their physical and intellectual Americanization—and had moved on to Americans of African descent; in the service of his ideas Hurston put in some time on a Harlem street corner with a pair of callipers, talking passersby into allowing her to measure their skulls.

Probably no one except her mother influenced Hurston more. Boas's fervent belief in the historic importance of African cultures had already had tremendous impact on W. E. B. Du Bois, who'd invited Boas to make the commencement address at Atlanta University in May 1906, when the city was increasingly in the grip of virulent anti-Negro, *Clansman*-inflamed outrage; there, Boas offered a clear counterstatement to Thomas Dixon's message of hate and shame, in a speech Du Bois called a revelation to himself and an inspiration to his students. Hurston was similarly inspired by the sense of importance that Boas gave to Southern black culture, not just as a source of entertaining stories but as the transmitted legacy of Africa—and as an independent cultural achievement in need of preservation and study. Boas literally turned Hurston around: he sent her back down South to put on paper the things that she'd always taken for granted. Furthermore, his sanction gave her confidence in the value of those things—the old familiar talk and byways—which was crucial to the sense of "racial health" and "easy self-acceptance" that so many relish in her work today. And still, it wasn't easy. Because, it seems safe to say, no black woman in America was ever simply allotted such strengths, no matter how bold she was or how uniformly black her hometown. They had to be won, and every victory was precarious.

As a child, Hurston informs us in her autobiography, she was confused by the talk of Negro equality and Negro superiority which she heard in the town all around her. "If it was so honorable and glorious to be black, why was it the yellow-skinned people among us had so much prestige?" Even in first grade, she saw the disparity: "The light-skinned children were always the angels, fairies and queens of school plays." She was not a light-skinned child, although her racial heritage was mixed. (Her mother was dark, her father notably fair; in *Jacob's Gourd Vine* she depicts him as the son of a plantation owner and a slave.) If the peculiarities of a segregated childhood spared Hurston the harshest brunt of white racism, the crippling consciousness of color in the black community and in the black soul was a subject she knew well and could not leave alone.

Such color-consciousness has a long history in African-American writing, starting with the first novel written by a black American, William Wells Brown's 1853 *Clotel:* a fantasy about Thomas Jefferson's gorgeous mulatto daughter, it took color prejudice "among the negroes themselves" as a premise. By the 1920s, the head of the NAACP was exhorting the Harlem community to strive against its "ever present race and color prejudice . . . against blacks," and the heroine of Wallace Thurman's bitterly funny novel *The Blacker the Berry* . . . was drenching her face with peroxide before going off to dance in Harlem's Renaissance Casino. But there is no more disconcertingly morbid document of this phenomenon than Hurston's prize-winning *Color Struck.* This brief, almost surreal play tracks a talented and very dark-skinned woman's decline into self-destructive madness, a result of her inability to believe that any man could love a woman so black. Although the intended lesson of *Color Struck* seems clear in the retelling, the play's fevered, hallucinatory vehemence suggests a far more complex response to color than Hurston's champions today can comfortably allow—a response not entirely under the author's control.

The play was published in 1925 in a magazine called *Fire!!,* which Hurston put together with Thurman and Langston Hughes, and which was meant to be more audaciously truthful than the larger journals of "New Negro" writing could afford to be. The writers put up the money themselves—and promptly lost it, snuffing *Fire!!* out after a single issue. Hurston alone had contributed two works; along with *Color Struck* she'd published a short story called "Sweat," one of the most powerful things she ever wrote and perhaps the harshest. This startling account of a marital war to the death is set in an entirely charmless Eatonville, in which no one is willing to protect the hardworking good-Christian laundress, Bertha, from the monstrous brutality of her philandering husband, Sykes. To stop him from beating her, Bertha has only one recourse, one meaningful threat: "Ah'm goin' tuh de white folks 'bout you."

It would be a mistake to say that "white folks" did not figure prominently in Hurston's early life, despite their physical remove. It was precisely because of the limited white presence that she took hold of racism not at its source but as it reverberated through the black community. But, just as important, whites around Eatonville were not the murderous tyrants of Richard Wright's Deep South childhood. If they exerted an equally powerful force on the imagination of a growing black writer, it was as tantalizing, world-withholding gods—and as a higher court (however unlikely) of personal justice.

There is a fairy-tale aspect to the whites who pass through Hurston's autobiography: the "white man of many acres and things" who chanced upon her birth and cut her umbilical cord with his knife; the strangers who would drive past her house in their "heavenly chariots" and give her rides out toward the horizon. (She had to walk back, and was invariably punished for her boldness.) Most important was a pair of white ladies who visited her school and were so impressed by her reading aloud—it was the myth of Persephone, crossing between realms of dark and light, which, she recalls, she read exceptionally well because it "exalted" her—that they made her a present of a hundred new pennies and the first real books she ever owned.

Hurston's autobiography, which won an award for race relations in 1943, has since been reviled by the very people who rescued her fiction from oblivion, and for the same reason that the fiction was once consigned there: a sense that she was putting on a song and dance for whites. In fact, there is nothing in *Dust Tracks on a Road* that is inconsistent with the generally romantic images of white judges and jurors and plantation owners which form a fundamental part of Hurston's most deeply admired work. The heroine of *Their Eyes Were Watching God* ends up on trial for the murder—in self-defense—of the man she loved, who lost his senses after being bitten by a rabid dog and came at her with a gun. The black folks who knew the couple side against her at the trial, hoping to see her hanged. It is the whites—the judge and jury and a group of women originally gathered just for curiosity's sake—who see into the anguished depths of a black woman's love and acknowledge her dignity and her innocence.

Does this reflect honest human complexity or racial confusion? In what world, if any, was Hurston ever at home? While at Barnard, she apparently told the anthropologist Melville Herskovits that, as he put it, she was "more white than Negro in her ancestry." On her first trip back South to gather evidence of her native culture, she could not be understood because of her Barnard intonations. She couldn't gain people's confidence; the locals claimed to have no idea what she wanted. When Hurston returned to New York, she and Boas agreed that a white person could have discovered as much.

So she learned, in effect, to pass for black. In the fall of 1927, in need of a patron, she offered her services to Mrs. R. Osgood Mason, a wealthy Park Avenue widow bent on saving Western culture from rigor mortis through her support of Negro artistic primitivism. (Mason already had Alain Locke—whose idea of primitivism was derived from Picasso—and Langston Hughes, then a budding Marxist, under her heavy white wing.) For several years, Hurston's "Godmother" paid for her to make forays to the South to collect Negro folk material. Her findings were not always as splendidly invigorating nor her attitude as positive as they later appeared. "I have changed my mind about the place," she wrote despairingly to Mason from Eatonville in a letter of 1932. "They steal everything here, even greens out of a garden." But she became increasingly accomplished at discovering what she had been hired to find, and the results (if not always objectively reliable) have proved an invaluable resource. Alan Lomax, who worked with Hurston on a seminal 1935 Library of Congress folk-music-recording expedition, wrote of her unique ability to win over the locals, since she "talks their language and can out-nigger any of them."

The fruits of her fieldwork appeared in various forms throughout the early thirties: stories, plays, musical revues, academic articles. Her research is almost as evident in the 1934 novel, *Jonah's Gourd Vine,* as in her book of folklore, *Mules and Men,* which appeared the following year. Now routinely saluted as the first history of black American folklore by a black author, *Mules and Men* begins—in what can only be a tribute to Boas—with a little tale in dialect that offers honor to the Jews as the people who, to their eternal pain and discomfort, have stolen the largest amount of soul from God. Hurston claimed to have known the story since her childhood; this unlikelihood is one reason to suspect that her anthropology is as touched with fiction as her fiction is touched with anthropology. But beyond the questioning of specific examples, *Mules and Men* was severely faulted by black critics of its time for its exclusion of certain nonfictional elements of the Southern Negro experience: exploitation, terror, misery, and bitterness.

By then, however, Hurston had won enough recognition to receive a Guggenheim grant to go off to study voodoo practices in the Caribbean. This was not a happy trip. The anecdotal study she produced—*Tell My Horse,* published in 1938—is tetchy and belligerent, its author disgusted by the virulent racism of light-skinned mulattoes toward blacks in Jamaica, and as distinctly put out by the unreliability and habitual lying she experienced among the Haitians. In any case, this particular trip had been prompted less by an interest in research than by a need to escape from New York, where she'd left the man she thought of as the love of her life—a still mysterious figure who belongs less to her biography than to her art. In a period of seven weeks, in Haiti in the fall of 1936, she wrote *Their Eyes Were Watching God,* a novel meant to "embalm all the tenderness of my passion for him."

In her autobiography, Hurston quickly dismisses her first marriage and entirely neglects to mention her second; each lasted only a matter of months. She wed her longtime Howard University boyfriend in May 1927 and bailed out that August. (Apparently unruffled, Hurston wrote her friends that her husband had been an obstacle and had held her back.) In 1939 her marriage to a twenty-three-year-old WPA playground worker dissolved with her claims that he drank and his claims that she'd failed to pay for his college education, as she had promised, and had threatened him with voodoo. "The great difficulty lies in trying to transpose last night's moment to a day which has no knowledge of it," she writes in *Dust Tracks on a Road.* She concludes, "I have come to know by experience that work is the nearest thing to happiness that I can find."

Those admirers who wish Hurston to be a model feminist as well as a racial symbol have seized on the issue of a woman's historic choice between love and work, and have claimed that Hurston instinctively took the less travelled path. On the basis of Hurston's public insouciance, Alice Walker describes, with delicious offhand aplomb, "the way she tended to marry or not marry men, but enjoyed them anyway, while never missing a beat in her work." No sweat, no tears—one for the girls. It is true that Hurston was never financially supported by a man—or by anyone except Mrs. Mason. Hemenway, her biographer, writes that it was precisely because of her desire to avoid "such encroachments" on her freedom that her marriages failed.

Without doubt, Hurston was a woman of strong character, and she went through life mostly alone. She burned sorrow and fear like fuel, to keep herself going. She made a point of not needing what she could not have: whites who avoided her company suffered their own loss; she claimed not to have "ever really wanted" her father's affection. Other needs were just as unwelcome. About love, she knew the way it could make a woman take "second place in her own life." Repeatedly, she fought the pull.

There is little insouciance in the way Hurston writes of the man she calls P.M.P. in *Dust Tracks on a Road.* He was "tall, dark brown, magnificently built," with "a fine mind and that intrigued me. . . . He stood on his own feet so firmly that he reared back." In fact, he was her "perfect" love—although he was only twenty-five or so to her forty, and he resented her career. It is hard to know whether his youth or his resentment or his perfection was the central problem. Resolved to "fight myself free from my obsession," she took little experimental trips away from him to see if she could bear it. When she found she couldn't, she left him for good.

Her diligent biographer, who located the man decades later, reports that he never understood exactly what had happened. She'd simply packed her bags and gone off to the Caribbean. Once there, she wrote a book in which a woman who has spent her life searching for passion finally finds it, lets herself go within its embrace, and learns that her lover is honest and true, and that she is not being played for a fool—despite the familiar fact that he is only twenty-five or so and she is forty. (He reassures her, "God made it so you spent yo' ole age first wid somebody else, and saved up yo' young girl days to spend wid me.") And then, in the midst of love's fulfillment, the woman is forced—by a hurricane and a mad dog and a higher fate—to shoot him dead, and return to just the state of enlightened solitude the author had chosen for herself.

Their Eyes Were Watching God brought a heartbeat and breath to all Hurston's years of research. Raising a folk culture to the heights of art, it fulfilled the Harlem Renaissance dream just a few years after it had been abandoned; Alain Locke himself complained that the novel failed to come to grips with the challenges of "social document fiction." The recent incarnation of Hurston's lyric drama as a black feminist textbook is touched with many ironies, not the least of which is the requirement that it be considered as social document fiction. The paramount ironies, however, are two: the heroine is not quite black, and becomes even less black as the story goes on; and the author offers one of the most serious Lawrentian visions ever penned by a woman of sexual love as the fundamental spring and power of life itself.

The heroine of *Eyes,* Janie Crawford, is raised by her grandmother, who grew up in "slavery time," and who looks on in horror as black women willingly give up their precious freedom for chains they forge themselves. "Dis love! Dat's just whut's got us uh pullin' and uh haulin' and sweatin' and doin' from can't see in de mornin' til can't see at night." But no one can give a woman what she will not claim. Nanny's immovable goal to see Janie "school out" meets its match in the teenager's bursting sexuality. Apprehensive, Nanny marries her off to a man with a house and sixty acres and a pone of fat on the back of his neck. "But Nanny, Ah wants to want him sometimes. Ah don't want him to do all de wantin'," Janie complains, and she walks off one day down the road, tossing her apron onto a bush.

It isn't exactly Nora slamming the door. There's another man in a buggy waiting for Janie, and another unhappy marriage—this time to a bully who won't let her join in the dazzling talk, the wildly spiralling stories, the fantastic games of an Eatonville that Hurston raises up now like a darktown Camelot. After his death, a full twenty years later, Janie is rather enjoying the first freedom of widowhood when a tall, laughing man enters the general store and asks her to play checkers: "She looked him over and got little thrills from every one of his good points. Those full, lazy eyes with the lashes curling sharply away like drawn scimitars. The lean, overpadded shoulders and narrow waist. Even nice!"

It's the checkers almost as much as the sex. After Nanny, this man, who is called Tea Cake ("Tea Cake! So you sweet as all dat?"), is the staunchest feminist in the novel. He pushes Janie to play the games, talk the talk, "have de nerve tuh say whut you mean." They get married and set off together to work in the Everglades, picking beans side by side all day and rolling dice and dancing to piano blues at night. Hurston isn't unaware of the harsh background to these lives—trucks come chugging through the mud carrying migrant workers, "people ugly from ignorance and broken from being poor"—but she's willing to leave further study to the Wrights and the Steinbecks. Her concern is with the flame that won't go out, the making of laughter out of nothing, the rhythm, the intensity of feeling that transcends it all.

During the 1970s, when *Their Eyes Were Watching God* was being rediscovered with high excitement, Janie Crawford was granted the status of "earliest . . . heroic black woman in the Afro-American literary tradition." But many impatient questions have since been asked about this new icon. Why doesn't Janie speak up sooner? Why can't she go off alone? Why is she always waiting for some man to show her the way? Apologies have been made for the difficulties of giving power and daring to a female character in 1936, but then, Scarlett O'Hara didn't fare too badly with the general public that year, and Hurston herself was hardly lacking in these qualities. The fact is that Janie was not made to suit independent-minded female specifications of any era. She is not a stand-in for her author but a creation meant to live out other possibilities, which are permitted her in large part because—unlike her author—she has no ambition except to live, and because she is beautiful.

"I got an overwhelming complex about my looks before I was grown," Hurston wrote her friend and editor Burroughs Mitchell in 1947, but went on to state that she had triumphed over it. "I don't care how homely I am now. I know that it doesn't really matter, and so my relations with others are easier." Despite the possible exaggerations of a momentary declaration, this vibrantly attractive woman was well acquainted with what might be called the aesthetic burdens of race ("as ugly as Cinderella's sisters" is a local expression meaning "Negro," Hurston reported from Florida to Mrs. Mason), and she spared her romantic heroine every one of them.

Janie recalls of an early photograph, "Ah couldn't recognize dat dark chile as me," and by the middle of the book neither can we. By then, we've heard a good deal about Janie's breasts and buttocks and so extraordinarily much about her "great rope of black hair"—a standard feature of the gorgeous literary mulatto, going back even to Clotel's *mother*—that one critic wrote that it seemed to be a separate character. But it is only when Janie and Tea Cake get to the Everglades and confront the singularly racist Mrs. Turner, eager to "class off" with other white-featured blacks ("Ah ain't got no flat nose and liver lips. Ah'm uh featured woman") that we hear of Janie's "coffee-and-cream complexion" and "Caucasian characteristics."

The transformation is both touching and embarrassing—something like George Eliot's suddenly making Dorothea sublimely beautiful in the Roman-museum scene of *Middlemarch.* It's as though the author could no longer withhold from her beloved creation the ultimate reward: Dorothea starts to look like a Madonna, and Janie starts to look white.

With Hurston, though, pride always rushes back in after a fall. These alternating emotional axes are what make her so unclassifiable, so easily susceptible to widely different readings, all of which she may intend. For Janie never acts white, or even seems to care whether she looks that way. She is sincerely mystified by Mrs. Turner's tirades. "We'se uh mingled people," she responds, seeming to rebuke her author's own reflexive notions of beauty, too. "How come you so against black?"

Notably, Tea Cake doesn't share this naive mystification at why a black person would be "so against black"—for he, unlike Janie, has experienced the boot-heel of white racism firsthand. We are given only a glimpse of such experience, late in the story, when the lovers have been propelled by the hurricane all the way to Palm Beach, and Tea Cake is conscripted, by white men with rifles, into a crew assigned to clear the wreckage and bury the dead. The gravedigging scene that follows is floatingly surreal. In answer to an order that the rotting bodies be separated by race, the crew—black and white—complains that many are so far gone no racial distinction can be made; the guards return with instructions to "look at their hair, when you can't tell no other way." Even the mass-burial ditches here must be segregated, with white remains given cheap pine coffins while blacks are sprinkled with lime and thrown directly in. "They's mighty particular how dese dead folks goes tuh judgment," Tea Cake observes to a fellow worker. "Look lak dey think God don't know nothin' bout de Jim Crow law." Hurston, having lightly but indelibly brushed in this supreme social derangement, moves on quickly; the entire scene takes a page and a half. Tea Cake then easily escapes, without any confrontation, and not in fear or protest but because "the thought of Janie worrying about him made him desperate." The story returns to the personal as swiftly as the lovers resolve to return to the Everglades, where at least, they remark, the white folks know them. "De ones de white man knows is nice colored folks," Janie agrees as they set off; "de ones he don't know is bad niggers."

Although Janie spends much of the book struggling to gain the right to speak her mind, she is not particularly notable for her eloquence; even in the above statement—unusually analytic—she is paraphrasing Tea Cake.

There is, however, a great deal of poetry of observation running through her head, which we hear not as her thoughts, precisely, but in the way the story is told. Those who ponder "narrative strategies" have pulped small forests trying to define Hurston's way of slipping in and out of a storytelling voice that sometimes belongs to Janie and sometimes doesn't and, by design, isn't always clear. (As in *Mrs. Dalloway*, the effect is of a woman's sensual dispersal through the world.) Janie's panting teenage sexuality is rendered in a self-consciously hyperadolescent prose of kissing bees and creaming blossoms—prose that Wright seized on for its "facile sensuality" and that Hurston's admirers now quote with dismaying regularity as an example of her literary art. But Hurston at her best is simple, light, lucid, or else, just as simply, biblically passionate. Janie wakes to see the sun rise: "He peeped up over the door sill of the world and made a little foolishness with red." (There is an archaic sense of power in Hurston's sexing of all things: "Havoc was there with her mouth wide open.") As for Tea Cake, even as Janie tries to push his image away, Hurston writes that he "seemed to be crushing scent out of the world with his footsteps. Crushing aromatic herbs with every step he took. Spices hung about him. He was a glance from God."

This is a sermon from the woman's church of Eros. And like the sermons in which Hurston was schooled—like her entire book, as it winds in and out of this realization of sexual grace—her message lives in its music. At her truest as a writer, Hurston was a musician. The delightfully quotable sayings that she "discovered" on her field trips (many of which appear as plucked examples in *Mules and Men* and her other books) are embedded in this single volume like folk tunes in Dvořák or Chopin: seamlessly, with beauties of invention often indistinguishable from beauties of discovery. The rhythms of talk in her poetry and the substance of poetry in her talk fuse into a radiant suspension. "He done taught me de maiden language all over," Janie says of Tea Cake, and there may be some truth to the tribute: Hurston had never written this way before, and she never rose to it again. It seems likely that without the intensity of her feelings for "P.M.P.," this famously independent woman would not have written the novel that is her highest achievement and her lasting legacy. It perhaps complicates the issue of a woman's life and work that the love she tore herself away from so that she could be free, and free to write, turned out to have been the Muse.

▪ ▪ ▪

Hurston's ability to write fiction appears to have dried up after the commercial failure of *Their Eyes Were Watching God,* which sank without a trace soon after publication. Her next novel, *Moses, Man of the Mountain,* published in 1939, reads like a failed reprise of the Bible-based all-Negro Broadway hit *Green Pastures,* with the story of Exodus as its black-face subject. ("Oh, er—Moses, did you ask about them Hebrews while you was knocking around in Egypt?") Gone is the miraculous ear. Gone, too, are her great humor and heart. *Moses* is a weary book, heavy with accumulated resentments; jealous arguments over shades of skin color abound, and one of the few themes that feel emotionally true is Moses' exasperation with his complaining, unappreciative people.

Hurston's disillusionment is as fully evident in her mordant, angry journalism of the 1940s. With a deliberate, almost revelling perversity, she witheringly commends the Southern custom of whites holding their own "pet Negroes" exempt from the acknowledged sins of the race (their eager pets may return the favor), and explains that this racial system would offer some mutual protection, at least, "if ever it came to the kind of violent showdown the orators hint at." She rails against the substandard Negro colleges she calls "begging joints," the kind of institution white liberals feel good about donating their money to, although they cheat black students out of any real, competitive education. The title of one article—"Negroes Without Self-Pity"—speaks for itself.

"I don't see life through the eyes of a Negro, but those of a person," she told the *New York World-Telegram* in an interview in 1943. The statement was perfectly in keeping with the contents of *Dust Tracks on a Road,* which had just won Hurston a thousand-dollar award for contributing to race relations. But the paper went on to quote her as saying—she would soon deny it all, but a reader of her journalism can well imagine the harsh, miscalculated irony that could have produced such words—that blacks living in the South were better off than their brethren in the North, that separate but equal facilities at least meant there were facilities, and that "in other words, the Jim Crow system works." Needless to say, such claims by the United States' most prominent black woman author—Hurston had just appeared on the cover of the *Saturday Review*—were seized on by the NAACP, and she was denounced by Roy Wilkins in the *Amsterdam News* for speaking "arrant and vicious nonsense" just to sell her book. Her claims to have been misquoted were neither widely reported nor widely believed. "I have become what I never wished to be," she wrote to an editor at the Associated

Negro Press, "a good hater." Actually, she was a terrible hater; the feeling nearly destroyed her.

In the next few years, two new novels were rejected, her poverty went from bohemian to chronic, and her health—although not her principles—gave way. In need of escape, she bought a houseboat and spent much of the mid-forties sailing Florida rivers: individualism, her refuge from racism, lapsed into nearly total isolation. She returned to New York in 1946, looking for work, and wound up in the campaign office of the Republican congressional candidate running against Adam Clayton Powell. When her side lost, she was stranded for a terrible winter in a room on 124th Street, in a far more frightening sort of isolation. She didn't ask for help, and she didn't get any. She felt herself slipping, surrounded by racist hostility, the whole city "a basement to Hell."

It was just after this that she wrote her last published novel, *Seraph on the Suwanee.* ("Suwanee" is the actual name of the river that Stephen Foster made famous in a more singable form.) The story of a white Southern woman and her family, the book contains no prominent black characters. Among Hurston's stalwart supporters, Alice Walker has called it "reactionary, static, shockingly misguided and timid," and Mary Helen Washington has called it "vacuous as a soap opera." Everyone agrees that Hurston had fallen into the common trap of believing that a real writer must be "universal"—that is to say, must write about whites—and that she had simply strayed too far from the sources that fed her. In fact, the book is poisonously fascinating, and suggests, rather, that she came too close.

The story of beautiful, golden-haired Arvay Henson, who believes herself ugly and unworthy of love, contains many echoes of Hurston's earlier work, but its most striking counterpart is the long-ago play *Color Struck.* The works set a beginning and an end to years of struggle with their shared essential theme—the destructive power of fear and bitterness in a woman's tortured psyche. Arvay is born to a poor-white "cracker" family; in a refraction of Hurston's own history, a preference for her older sister "had done something to Arvay's soul across the years." She falls in love with a magnificent fallen aristocrat, who rapes her—for Arvay this is an act of ecstatic, binding possession—and marries her. Feeling ever more convinced of his innate superiority, however, and tormented by her failure to live up to his perfection, she comes to hate him almost as much as she hates herself.

The book is a choking mixture of cynicism and compulsion. Hurston was desperate for a success, and hoped for a movie sale—hence, no doubt, the formulaic rape and the book's mawkish ending, in which Arvay learns to sing happily in her marital chains. But to reach this peace Arvay must admit, after years of pretense, that she is not really proud of her own miserably poor and uneducated family, that poverty and ignorance lend them neither moral stature nor charm, and that she is, in fact, shamed and disgusted by them. Arvay's last attempt to go home to her own people results in her burning down the house in which she was raised.

The book was sharply criticized because Hurston's white Southerners speak no differently from the Eatonville blacks of her earlier work. The inflections, the rhythms, the actual expressions that had been declared examples of a distinctive black culture were all now simply transferred to white mouths. The incongruous effects, as in her *Moses* book, point to a failure of technique, an aural exhaustion. But in a letter to her editor Hurston gave an even more dispiriting explanation for what she'd done. "I think that it should be pointed out that what is known as Negro dialect in the South is no such thing," she wrote in a repudiation nearly as sweeping as Arvay's, at once laying waste to her professional past and her extraordinary personal achievement. The qualities of Southern speech—black and white alike, she claimed—were a relic of the Elizabethan past preserved by Southern whites in their own closed and static society. "They did *not* get it from the Negroes. The Africans coming to America got it from them."

The novel's publication, in the fall of 1948, was swallowed up in a court case that tested all of Hurston's remaining capacity for resisting bitterness. That September in New York, an emotionally disturbed ten-year-old boy accused her of sexual molestation. The Children's Society filed charges, and Hurston was arrested and indicted. Although the case was eventually thrown out, a court employee spilled the news to one of the city's black newspapers—the white papers were presumably not interested—and the lurid story made headlines. Betrayed, humiliated, sick, Hurston contemplated suicide, but slowly came back to herself on a long sailing trip.

She never returned to New York. For the rest of her life, she lived in Florida, on scant money and whatever dignity she was able to salvage. In Miami, she worked as a maid. Later, she moved to a cabin up the coast that rented for five dollars a week, where she was devoted to her garden and grew much of her own food. She labored over several books, none considered publishable. Her radical independence was more than ever reflected in her politics: fervently anti-Communist, officially Republican, resisting anything that smacked of special pleading. When *Brown v. Board of Education* was decided, in 1954, she was furious—and wrote furiously, in a local newspaper—over the implication that blacks could learn only when seated next to whites, or that anyone white should be forced to sit beside someone black. It was plain "insulting." Although there was some hard wisdom in her conclusion—"the next ten years would be better spent in appointing truant officers and looking after conditions in the homes from which the children come"—her defiant segregationist position was happily taken up by whites of the same persuasion. Her reputation as a traitor to her people overshadowed and outlasted her reasoning, her works, and her life.

Hurston died in January 1960, in the Saint Lucie County welfare home in Fort Pierce, Florida, four days before the first sit-in took place at a Woolworth's lunch counter in Greensboro, North Carolina. She was buried in an unmarked grave in a segregated cemetery in Fort Pierce. All her books were out of print. In 1971, in one of the first important reconsiderations of writers of the Harlem Renaissance, the critic Darwin Tuner wrote that Hurston's relative anonymity was understandable, for, despite her skills, she had never been more than a "wandering minstrel." He went on to say that it was "eccentric but perhaps appropriate"—one must pause over the choice of words—for her "to return to Florida to take a job as a cook and maid for a white family and to die in poverty." There was a certain justice in these actions, he declared, in that "she had returned to the level of life which she proposed for her people."

The gleaming two-volume Library of America edition of Hurston's *Novels & Stories* and *Folklore, Memoirs, & Other Writings* makes for a different kind of justice. These books bring Hurston a long way from the smudged photocopies that used to circulate, like samizdat, at academic conventions, and usher her into the national literary canon in highly respectable hardback. She is the fourth African-American to be published in this august series, and the fifth woman, and the first writer who happens to be both. Although the Hurston revival may have been driven in part by her official double-victim status—a possibility that many will take as a sign that her literary status has been inflated—*Their Eyes Were Watching God* can stand unsupported in any company. Our chief canon-keeper, Harold Bloom, has written of Hurston as continuing in the line of the Wife of Bath and Falstaff and Whitman, as a figure of outrageous vitality, fulfilling the Nietzschean charge that we try to live as though it were always morning.

Outside of fiction, this kind of strength is mainly a matter of determination. For many who have embodied it in literature—Nietzsche, Whitman, Lawrence, Hurston—it is a passionate dream of health (dreamed while the simply healthy are sound asleep) which stirs a rare insistence and bravado. "Sometimes, I feel discriminated against, but it does not make me angry," Hurston wrote in 1928. "It merely astonishes me. How *can* any deny themselves the pleasure of my company?" In the venerable African-American game of "the dozens," the players hurl monstrous insults back and forth as they try to rip each other apart with words. (Both Hurston and Wright summon up the game, and quote the same now rather quaint chant of abuse: "Yo' mama don't wear no DRAWS, Ah seen her when she took 'em OFF"). The near Darwinian purpose was to get so strong that no matter what you heard about whomever you loved, you would not let on that you cared to do anything but laugh. It's a game that Richard Wright must have lost every time. But Zora Neale Hurston was the champ.

It is important not to blink at what she had to face and how it made her feel. Envy, fury, confusion, desire to escape: there is no wonder in it. We know too well the world she came from. It is the world she rebuilt out of words and the extraordinary song of the words themselves—about love and picking beans and fighting through hurricanes—that have given us something entirely new. And who is to say that this is not a political achievement? Early in *Their Eyes Were Watching God* Hurston describes a gathering of the folks of Eatonville on their porches at sundown: "It was the time to hear things and talk. These sitters had been tongueless, earless, eyeless conveniences all day long. Mules and other brutes had occupied their skins. But now, the sun and the bossman were gone, so the skins felt powerful and human. They became lords of sounds and lesser things. They passed notions through their mouths. They sat in judgment."

The powerless become lords of sounds, the dispossessed rule all creation with their tongues. Language is not a small victory. It was out of this last, irreducible possession that the Jews made a counterworld of words, the Irish vanquished England, and Russian poetry bloomed thick over Stalin's burial grounds. And in a single book one woman managed to suggest what another such heroic tradition, rising out of American slavery, might have been—a literature as profound and original as the spirituals. There is the sense of a long, ghostly procession behind Hurston: what might have existed if only more of the words and stories had been written down decades earlier, if only Phillis Wheatley had not tried to write like Alexander Pope, if only literate slaves and their generations of children had not felt pressed to prove their claim to the sworn civilities. She had to try to make up for all of this, and more. If out of broken bits of talk and memory she pieced together something that may once have existed, out of will and desire she added what never was. Hurston created a myth that has been gratefully mistaken for history, and in which she herself plays a mythic role—a myth about a time and place fair enough, funny enough, unbitter enough, glad enough to have produced a woman black and truly free.

A Perfect Lady

Eudora Welty

When Henry Miller set off to discover America in October 1940, there were several outstanding natives he was hoping to meet: Margaret Mitchell, Zora Neale Hurston, Walt Disney, Ernest Hemingway, and a little-known writer named Eudora Welty. Despite the alarmingly forward letter of introduction Miller had sent her sometime before, Welty—unfailingly courteous—received him as an honored guest. For three days, she drove Miller around the sights and surrounds of her native Jackson, Mississippi, the city where, at thirty-one, she lived with her widowed mother in a large Tudor-style house that her father had built. Welty's greeting was not only gracious but bold, since her mother refused to let Miller into the house—not because of his books but because of the letter, in which he'd offered to put Welty in touch with "an unfailing pornographic market" for her talents. In the extensive touring plans that Welty had devised for her exotic visitor, she arranged for at least two male chaperons to accompany her wherever they went.

The only shocking aspect of Miller's behavior, though, turned out to be his stupendous lack of interest in Southern history: he refused to take off his hat on a picnic at the local ruined plantation, and his apathy reached the point where he wouldn't turn his head to look out the car window. But if Welty had reason to dismiss this vagabond libertine as "the most boring businessman you can imagine," the contrast she offered to her own writing was hardly less extreme. Although she had published only a handful of stories, Welty already flaunted a distinct and not unshocking literary manner: deadly honest, ruthlessly funny, and as subversive of complacent American normalcy as that of any jaundiced Left Bank expatriate. At roughly the time she was boring Miller stiff-necked, editors of the *Atlantic Monthly* were worriedly censoring Welty's inspired black-jazz improvisation, "Powerhouse," and trying to explain to the genteel Southern-lady author why her story could not conclude with the slyly dirty lyrics—could she have had any idea what was being suggested?—to "Hold Tight, I Want Some Seafood, Mama" ("fooly racky sacky want some seafood, Mama!").

Welty knew that she was writing "something new," and she didn't expect success to come without a struggle. She had an uncannily keen ear, and she tended to write quickly, in bursts of energy that required little or no revision. The uproariously surging "Powerhouse" was completed in a single sitting, after she'd come home one night from a Fats Waller concert and felt an urge to spin the music's high-flying, improvisatory riffs out into words. Another story, called "Petrified Man"—in which the horror of a carnival freak show is easily outdone by the horror of a small-town beauty parlor ("this den of curling fluid and henna packs")—was rejected so many times that she burned it in disgust, and then rewrote it from memory when Robert Penn Warren, editor of the *Southern Review*, changed his mind and wanted it back. Its second appearance was in "O. Henry Prize Stories" of 1939.

Welty had a notably vivid sympathy for the freak and the grotesque, for the pygmy and the pickled and the blinkingly dim, characters who mostly served to set the wider population of her stories at a disadvantage. She knew her outsiders and she understood what people used them for. ("He's turning to stone," Leota the beautician sagely observes of the Petrified Man. "How'd you like to be married to a guy like that?") It is the uncertainly educated denizens of Welty's ingrown, posthistoric, Coca-Cola-sodden South who are more truly akin to monsters—albeit, at Welty's dismaying best, entirely guileless and extremely funny ones. The pitch-perfect talk she puts in their mouths seems to render moral judgment weightless; it is present but invisible, rising up like gas. But then, these thankless souls don't do any real harm—there is no Faulknerian bloodlust here—except in a way that Welty takes in glancingly, with a sharp little stab at the edge of her vision, as when the dramatically departing narrator of "Why I Live at the P.O." drags all her belongings onto the family porch:

> There was a nigger girl going along on a little wagon right in front.
>
> "Nigger girl," I says, "come help me haul these things down the hill, I'm going to live in the post office."
>
> Took her nine trips in her express wagon. Uncle Rondo came out on the porch and threw her a nickel.

Everyone knows who the real outsiders are in this world.

Welty published her first story in a small magazine called *Manuscript* in 1936, when she was twenty-six. In the decades since then, the young author of blithely daring fictions has become a monument, the Pallas Athena of Jackson. She has received the Pulitzer Prize and the Presidential Medal of Freedom; the Municipal Library of Jackson has been renamed the Eudora Welty Library; and the governor of Mississippi once decided that her birthday merited a statewide holiday. Welty's collected works have recently been published in two stout volumes by the Library of America, but she has already entered the national pantheon as a kind of favorite literary aunt—a living exemplar of the best that a quaint and disappearing Southern society still has to offer.

In her best-selling memoir, *One Writer's Beginnings,* Welty attributed her gifts and her success largely to having been the child of wonderful parents—an Ohio-born insurance man and a strong-minded West Virginia schoolteacher, who moved to Jackson after their marriage, in 1904. Welty, born in 1909, grew up in a prosperous home near the state capitol; she and her two younger brothers would roller-skate straight through the rotunda, part of a perfect idyll of childhood she portrayed, all kites and ice cream. She attended whites-only schools, of course—Richard Wright went to school in a very different Jackson during many of the same years—and the rest of her world was equally enclosed; the only black people she appears to have seen were contented servants. According to her own report, quite unremarkably she questioned nothing.

And yet by sixteen she was ready to get out. She convinced her parents that she was old enough to go away to college—first to Columbus, Mississippi, and then to Madison, Wisconsin, which was far enough from home, but in the wrong direction. After graduation, she moved to New York to attend Columbia Business School; this was 1930, and the theatre and Harlem jazz clubs and Martha Graham occupied her far more than her classes did. The life suited her: she returned to Jackson only when her father was dying, in 1931. Two years later she was back in New York, but financial worries and pressure from her mother brought her home again. It was then that she started to take photographs, principally in Jackson's black neighborhoods, where she went to buy jazz records.

Her real awakening, though, came with a job as a publicity agent for the Works Progress Administration, in 1935, for which she travelled by car or bus through the depths of Mississippi, and saw poverty—black and white—that she had never imagined before. Taking pictures now became her passion, and Welty published photographs before she published her first story. When the WPA job was over, in 1936, she returned to New York several times, searching for a job in publishing and pounding the pavement with her photographs. All she ever got was a small exhibition in an optician's store on Madison Avenue. Her subject was black Mississippians, in the fields or on the streets or simply looking outward, meeting impossible odds with a frank and powerful dignity.

It is telling that through the late thirties Welty tried to publish her stories and her photographs in a single volume. The impetus for what she knew to be her first genuine writing had come from the same shock of discovery—from her WPA travels, when, as she put it, "my feelings were engaged by the outside world, I think for the first time." The evidence of this experience is sometimes stark, as in "The Whistle," a story about impoverished tomato farmers who strip off their only warm clothing to cover the delicate crops during a frost, and in "A Worn Path," about an ancient black woman who undertakes a long journey on foot to get medicine for her grandson (and who serenely transcends the petty insults of the white people she meets along the way; she, too, receives a nickel). Despite the subjects, there is nothing didactic in these stories; Welty's tone remains as light and precise here as in her freak-show comedies. And like the comedies, these stories do not need to name the big subjects they touch on—race, deprivation, ignorance, morality—because the author's quick chiselling of character includes them all.

▪ ▪ ▪

A brilliant array of these stories appeared in Welty's first collection, *A Curtain of Green,* which was published late in 1941, after being held up for months while Welty trembled to ask the magisterial Katherine Anne Porter to complete a promised introduction. The book contained six years of varied, jolting, intense storytelling by an author who now gave no sign of ever having willingly stepped off her front porch. Welty had long been apologetic about not having "been in jail or trodden grapes like other young people"—as she put it in an autobiographical note to her first published story—but Porter's introduction transformed her into a holy vestal of inexperience, with no personal history to mention and no use for the wider world, a Little Miss Muffet entirely content at home, "where she lives with her mother, among her lifelong friends and acquaintances, quite simply and amiably." In part, this reflects Porter's amused condescension toward the comparatively awkward young woman whose work she admired. (Her famous reply to Welty's confession that she was still a virgin was "Yes, dear, and you always will be.") But if the image was rather unfair—to Welty's past, to her work and ambitions—it was one that Welty herself now seemed set on making real.

She had recently completed a book that accords all too well with this encroaching sense of her limits. *The Robber Bridegroom* is a novella-length fairy tale, mixing wicked stepmothers and golden heroines from the Brothers Grimm with old Mississippi legends of robbers and heroes along the Natchez Trace. Aiming at "once upon a time," Welty willingly threw off her gifts for character and for the revelations of ordinary talk and moved into a realm of symbolism, cutout figures, and lockjaw whimsy ("Ho! Ho! Ho!" said the second traveler . . ."). The reviews, however, were sometimes better than they had been for Welty's stories, and one critic—presumably noting that the plot revolved around a wandering innocent roughly educated by the world—compared the book to *Candide.* As an indication of Welty's theme, the comparison is not far off. But if she had sunk all her virtues into a parable, what was it about?

Of this, no one seemed to have a clue. Yet despite the book's tone of "joyous idiocy"—as another reviewer wrote, apparently in praise—it is clear that politics, in the broadest sense, runs like a murky source beneath the overbright tale. For Welty's "innocent" man—she uses the adjective repeatedly—is a Southern planter who accumulates great wealth and lands without any effort or desire. A purely noble gentleman, he is pushed on by the greed of his second wife—the evil stepmother—a spiritual foreigner from Kentucky who demands that he build her a vast plantation house (by its description, the very one that Henry Miller refused to salute). The book posits a golden age of the gracious South: a time when there were individual slaves but which was somehow distinct from the later age of institutionalized slavery. No wonder Welty had to write this as a fairy tale.

At the story's end, the wearied hero retreats from the scene, in a confusing speech of Yeatsian prophecy and horror, involving gold and slaves and buzzards and a planter who "makes a gesture of abundance with his riding whip." A gesture of abundance? Welty was clearly struggling with some peculiarly knotted ideas. In 1944 she wrote an extravagantly lyric essay entitled "Some Notes on River Country," in which she discussed the history of a lengthy section of the Mississippi: "Wonderful things have come down the current of this river. . . . Every kind of treasure, every kind of bearer of treasure has come down, and armadas and flotillas, and the most frivolous of things, too, and the most pleasure-giving of people." In this Watteau-like panorama, there is never any unfrivolous human cargo coming down the river; the only mention of slavery is in an episode of revenge taken by the French against the Indians. Possibly these evasions and glamorizations did not seem far-fetched to the average white resident of Jackson in the 1940s. But nothing about Welty's writing had seemed average before.

Little is known about Welty's life beyond what she has chosen to tell us. She continues to maintain that she has no personal history to speak of, and adds that whatever evidence exists to the contrary she is going to burn. Everything that matters has been in front of us all along, in the stories, the novels, and the literary and autobiographical essays. Nevertheless, despite the evident truth of this declaration of literary values, Welty's collected works may now be read alongside the first biography of the author ever attempted—needless to say, without her cooperation. Ann Waldron, the persistent author of *Eudora: A Writer's Life*, details several early trips to Welty's door, and a response as intractable as it was—what else?—gracious; rather like the response to Henry Miller's asking her to write pornography. But it took Waldron a lot longer than it did Miller to realize that Welty was not going to produce the goods.

With Welty's friends instructed not to talk, and with no access to private papers, Waldron depends on a lot of previously published material, frankly throws up her hands at the gaps—her opening chapter is titled "The Teenager"—and discovers a few simple but resonating facts. It is appalling that just about everyone willing to speak of Welty in her youth refers to her physical unattractiveness: "She was ugly to the point of being grotesque," an unnamed source reports in what soon amounts to a chorus. And while one would gladly do violence to a former belle who tells Waldron "I was pretty, so our paths didn't cross much," her statement does make clear what kind of segregation must have been the ruling evil of Welty's early life.

Equally important is how she overcame it. Almost every statement about Welty's looks comes to a "but" and turns itself around: but she was so nice, she was so helpful, she was so enthusiastic that her looks entirely ceased to matter. She did not have dates, but she never minded helping other girls get dressed for theirs. This was selflessness as survival, graciousness learned by the rack and the screw. Add to it a forcefully dominating mother and a fearful but hungry spirit, and what do you get? A woman who felt honor-bound to return home at twenty-two, when her mother was widowed, and then spent the next quarter century stealing as many weeks and months away as possible—she had to go to New York to get some privacy, she frets in a letter, she had to go to get some time to write—while fulfilling the obligations of an almost egregiously dutiful daughter. (In 1952 Katherine Anne Porter visited Jackson and was furious when she learned that Welty—aged forty-two—had to ask her mother if Porter could come for dinner; what's more, her mother said no.) You also get, it seems, a woman who spent seventeen years of her life in love with a homosexual man.

"Everybody is asking about John Robinson these days," Welty complained during an interview in 1993. The role that Waldron gives him is indeed central—and justifiably so, in the light of his significance not only in Welty's life but for her work. A tall and handsome man and an aspiring writer, Robinson became a friend of Welty's in 1933, when he returned to Jackson from graduate school. Although he moved to New Orleans in 1936, their companionship flourished; after he joined the army, in 1941, Welty rushed to meet him whenever he was on leave. In these years, Welty's letters to the multiply married Katherine Anne Porter are a bit breathless about the quantities of camellias that he brought her but reveal little of unbotanical interest. Although Waldron concludes that the available facts do not establish "whether they were friends or lovers," she notes that Welty's fiction is fraught with signs that, sexually, something was horribly wrong.

The books and stories would be quite sexless, in fact, were it not for the rapes—which are among the strangest such scenes in modern fiction. Without violence, without emotion, even without flesh, these peculiar events are all abstraction and tortured syntax. He "robbed her of that which he had left her the day before," Welty writes in *The Robber Bridegroom.* (What the robber took before were her clothes.) "He violated her and still he was without care," according to a 1942 story called "At the Landing," in which the heroine eats an enormous meal immediately afterward, gratefully demonstrating her "now lost starvation." Waldron convincingly suggests a source for this dire but bodiless sex in Welty's deep entanglement with a man who would not touch her. But the biographer ignores the far more meaningful influence that Robinson seems to have had on Welty's writing—an influence that was profoundly political.

The son of an old and wealthy Delta family, Robinson was part of the historic South in a way that Welty never was. His stepgrandfather was Mississippi's infamous Governor J. K. Vardaman, the racist demagogue responsible for the Jim Crow laws passed after his election, in 1904—the very year when Welty's parents moved to Jackson. Robinson seems to have been an impassioned dilettante of local history: an explorer of the big old houses and the little river towns (her "River Country" essay was sent to him for his approval) and a scholar of family histories. Perhaps his enthusiasm was irresistible; perhaps she wanted more than anything to please him; perhaps—particularly after her first book—she simply needed a new subject. But Welty's abrupt about-face from the roads of the WPA to the legendary trails of the Natchez Trace—and then on to the plantations of the Delta—would probably not have happened without him.

From 1941 to 1942 she completed seven stories of the Natchez Trace for a new collection, *The Wide Net;* the title story was dedicated to Robinson. Published in 1943 to disappointed reviews, the stories are often artificial and overliterary, the work of a gifted writer clearly struggling with the burden of being an Author. At Welty's best, however, her artifice takes on an intensely lyric, dancing energy—animating characters as simple in outline and rich in color as commedia dell'arte figures—that keeps the action hovering just above reality. One story that manages to remain aloft is the iridescent "Asphodel," in which three gossipy old maids confront the buck-naked god Pan while on a picnic at—where else?—the local ruined plantation, and run off, pursued by a pack of goats. The story seems part mock-Faulkner and—in the fleet departure of the suddenly sexually energized ladies—part Martha Graham, and suggests Welty's ability to slip quietly into place among these far more thunderously grand American mythologists.

The inspiration for her first novel, *Delta Wedding,* was owed directly to John Robinson. In 1945 he suggested that Welty read the diaries of his great-great-grandmother, a Delta plantation mistress who recorded the events of her life from 1832 through 1870. The work that grew from this chronicle was a labor of love—in some ways literally, with Robinson reading each chapter as it was completed. Because Welty didn't feel comfortable writing about a period she'd never known, the setting was moved to 1923—a year she chose precisely because it offered "nothing except the family" to concentrate on. The shift took her as far as possible from any social considerations and also brought her into the period of a literary style that displays the unmistakeable influence of Virginia Woolf. (*Mrs. Dalloway* is set on an entirely unextraordinary day in 1923.) Despite the appearance of an occasional automobile, however, or a reference to baying dogs at nearby Parchman prison—the lowest hell of the Jim Crow system, built by Governor Vardaman only in 1905—one could read the entire book without suspecting the transposition in time, or even realizing that the Civil War had occurred. Its characters, the Fairchild family of Shellmound, are benevolent, if temperamental, aristocrats, and their empire is maintained by largely comic and apparently contented Negroes—Bitsy, Roxy, Little Uncle, Vi'let, Man-Son—who give little sign of wanting the world any other way.

When *Delta Wedding* was published in 1946, critical swoons over Welty's language were mixed with horror at the book's contents. In *The Nation,* Diana Trilling wrote almost wistfully of the rich material bountifulness of these characters' lives, but drew herself up to conclude, point-blank, "This is a value system to which I deeply oppose myself"; *Time* called it all that could be expected from a member of the Junior League. Ironically, the very timelessness of Welty's book emphasized a point that this country—in the wake of Wright's *Black Boy,* a book Welty refused to review—was just beginning to address: namely, that in the twentieth century, parts of the South had effectively re-created conditions little better than those that had prevailed before the firing on Fort Sumter. ("This was the culture from which I sprang," Wright's 1945 book concluded. "This was the terror from which I fled.")

Certainly this was true of the Delta; studies from the early forties found that it stood alone in its brutal restrictions on Negro life. When, in *Delta Wedding,* the field hand Man-Son raises his hat to the Fairchilds' spirited, Scarlett O'Hara–like daughter, Dabny, what seems an act of devotion is also (unbeknownst to the reader) an act of obedience to what passed locally for law, not only in 1923 but in 1946. Welty claimed that she did not condone or endorse—this was not the job of the novelist—but merely described. What she describes, however, so gorgeously, is a dreamscape, a never-never land:

> Above in an unbroken circle, all around the wheel of the level world, lay silvery-blue clouds whose edges melted and changed into the pink and blue of sky. Girls and horses lifted their heads like swimmers. Here and there and far away the cotton wagons, of hand-painted green, stood up to their wheel tops in the white and were loaded with white, like cloud wagons. All along, the Negroes would lift up and smile glaringly and pump their arms—they knew Miss Dabny was going to step off Saturday with Mr. Troy.

This is not the work of a camera; it is, rather, a familiar myth turned out in a new, magnificent style. *Delta Wedding* might have been the greatest plantation novel of all had Welty had the courage of Margaret Mitchell's convictions, or else the courage to expose the lies and fears that lay beneath them. Instead, she spread fairy dust over the cotton fields and refused to confront or even explore any of her pretty characters. She couldn't afford to. The great literary model of family rapture that Welty adopted for her treatment of the Fairchilds, Woolf's *To the Lighthouse,* offered ample rationale for authorial subjectivity—for what Welty called, in a later appreciation of Woolf's masterpiece, a book written without "a shadow of detachment," one that seems to become a part of the substance it describes. But in *Delta Wedding* the Woolfian style of shifting consciousness and rhythmic undertow is used not to expose the flickering correspondence between inner and outer realities (the merging of which, in Woolf, makes up the actual substance described) but to avoid the existence of inner reality altogether. The main observers in Welty's story are outsiders—a visiting Jackson cousin, and a matriarch from Virginia who admits she will never know the "story" inside her Delta family's heads; the mysteries they witness and the words they hear do not connect to any human impulses they can explain or understand. As a result, *Delta Wedding* is a tour de force of distraction—food, flowers, dotty old aunts—always shifting focus around a nonexistent core.

Perhaps this emphasis on surface justly reflects how such a family lived with itself, but the artistic price Welty paid for so successfully mirroring its self-deception is heavy. The dazzled yet increasingly impatient reader does not ask that the author step outside her characters or proffer her disapproval—Welty never did that, even when describing characters of no moral grace whatever, and still she could make us smile through our horror—but to open her eyes, to stop squinting in the glare of her Delta sunshine and her equally blinding prose. But this is the book in which Welty began to turn her attentions from people to what she later termed "the magic of place." The only human moments to fully break through her cloud-wagon pastorale are a sardonic sketch of a black woman servant at home in her quarters, and several acutely felt passages in which various (white) women muse on their loneliness and their romantic longing. Beyond this, her characters remain blurred, as in a faded photograph—closed off in their delusions, small and indistinct against the beauties of the land that cannot save them or this book from being, finally, empty and unfulfilled.

In the fall of 1946, Welty used her *Delta* money to follow Robinson to San Francisco, where he had moved after leaving the army. She spent five months living alone in cheap single rooms, writing stories and nursing Robinson through his colds and depressions, then returned to Jackson in the spring. It was during these months that she finished what she has always called the most personally meaningful among her stories, "June Recital," which became the centerpiece of her next collection, *The Golden Apples*. A Willa Catheresque story—one can now feel Welty trying out models—about artistic aspiration and personal failure, it tells of an old-maid piano teacher who is slowly driven insane; rumored to have murdered her overbearing mother, she is led away after trying to burn down their house.

And that may not be the worst of the fates represented in this collection, which traces the lives of several residents of fictional Morgana, Mississippi, through a series of related stories—a device that allowed Welty to approach the dimensions of a novel while sidestepping the difficult requirements of the longer form. The opening story, "Shower of Gold," is, in fact, one of Welty's one-day-production wonders—the quicker she spilled it out, often, the better it was—a compressed but full-blown tragicomedy concerning marital infidelity and aching loss and tiny, terrifying twins on skates. In these early years of Morgana, anchored in time by the inauguration of Governor Vardaman, Welty's poor-white country town is a place where satyrs occasionally spring out from behind the trees. As with "Asphodel"—as with *Delta Wedding*—Welty aspires to a precarious balance between reality and fantasy; the balance does not hold for long, yet as the stories subside into an ever-flatter realism it seems as though the South itself were running out of myth.

In terms of language, the pungent folksiness of a local talker's confiding account—always Welty's uncanniest accomplishment—gives way to a more standard authorial omniscience, and it is not a single voice but an entire social atmosphere that resounds in simple but startling narrative observations like "Tonight, it was only the niggers, fishing" or "Cat, the niggers' cat, was sunning on a post," both from a story about budding sexual curiosity and fear among girls at summer camp called "Moon Lake." Casually insinuating, the word "nigger" recurs in many such quiet but charged ways throughout this collection. It emerges from the mouth of an endearingly kind white woman when she talks about a man she rather respects—"a real trustworthy nigger," Mrs. Rainey says, vouching for Old Prez's contribution to the tale she is telling. Another story contains the revelation that the old-maid piano teacher's tragedy began years before, when "a crazy nigger had jumped out of the school hedge" and raped her—"pulled her down and threatened to kill her" is how Welty puts it, avoiding the word "rape" just as the town does. The non-Southern reader soon comes to feel that the word "nigger" was in such common use that it lacked any specifically intended negative meaning or emotional charge (no matter what Northerners or blacks might think)—no nickels are thrown in this book, and the teacher's rape is presented as a dully unelaborated matter of fact. The result, cumulative in effect, is that individual characters and even communities are absolved of any conscious racial malignity of purpose; and, conversely, that the burden of racism seems all the more impossible to grasp and lift.

And then, after all this, Welty's final story, "The Wanderers," sets us down in a radically transformed, truck-route and sawmill Morgana, a part of the New South, and "nigger" is consistently replaced on the page by "Negro"—with the single exception of one lowlife old man's tale of long-ago goings-on. The change in language clearly had nothing to do with an access of political consciousness on Welty's part: "The Wanderers" was published before "Moon Lake"—March 1949 in *Harper's Bazaar,* as opposed to Summer 1949 in *The Sewanee Review.* It is possible, of course, that *Harper's Bazaar* simply refused to publish the story any other way. The new locution is stiffly used and not always less discomfiting than the old (in a motif reworked years later by Raymond Carver, a widow is said to have kept her husband's dying breath in a toy balloon, "until a Negro stole it"). The effect, perhaps inadvertent, is to suggest the clumsiness of the South's slow change from Old to New, and to make it clear that some things had still not changed at all.

It is part of Welty's engaging complexity that, despite her obvious affection for Morgana, the town turns out to be stultifying for the more sentient whites who live there; many of her characters have the sense to dream, at least, of getting out. Perhaps the greatest tragedy is that of Mrs. Rainey's daughter, Virgie, a wonderfully fearless little girl who "caught lightning bugs and tore out their lights for jewelry" and became the town's prize piano student and who did leave, once, to go out into the world. We never find out how she failed, what caused her to retreat—writing separate stories rather than a novel allowed Welty to leave this crucial matter out. But we do get an oblique hint at one possible reason. In "The Wanderers" we are confronted with Virgie back in Morgana, past forty and overdressed and still unmarried (" 'Don't put it off any longer,' " an old acquaintance warns, "grimacing out of the iron mask of the married lady"), and now somewhat numb with what may or may not be grief over the fact that she has just buried her mother—the sainted woman who'd stood and waited every single evening in their yard or their house for Virgie to come home.

Welty made a second trip to San Francisco in the fall of 1947 and then went back to Jackson. It was New York the next year—she tried to break into writing for the theatre—and then back to Jackson. In 1949 she travelled to Europe, apparently to meet up with Robinson in Florence, where he was studying on a Fulbright grant. (His letters of recommendation had included Welty and ex-Governor Vardaman.) And it was there that Robinson fell in love with the man he was to live with for the rest of his life. Welty told him how happy she was for him, and their friendship suffered no great break. Nor was there any apparent effect upon her work, except that when she wrote of herself in fiction again, it was not as an old maid but as a widow.

Alone, permanently now, she travelled more than ever, but the effect on her writing was only a deeper drawing-in. Welty's stories from the fifties—set in San Francisco or Ireland or on a boat to Naples—seem thin and conventionalized, the experience never quite convincingly firsthand. Yet her treatment of the contemporary South comes to seem even less credible, not by reason of defects in her talents but by her design. This is not a new development. *The Ponder Heart,* a 1954 novella, is essentially *The Robber Bridegroom* rewritten, a heavily whimsical fable about another purely innocent Southern man—Uncle Daniel Ponder, so rich and amiably crazy he gives all his possessions away. The major difference is that Welty's literary model is no longer the Brothers Grimm or Yeats but something akin to the folksy humbug of Will Rogers, and that this time she had a great success.

The slender volume became an alternate selection of the Book-of-the-Month Club and fulfilled Welty's theatrical dream by opening in a stage adaptation on Broadway in 1956. *Life* magazine ran a feature on the premiere of the dizzy new comedy in the same issue that it ran Faulkner's "A Letter to the North," a directive warning the country to go slow in forcing changes on the South in the increasingly volatile matter of civil rights. ("They don't mean go slow," Thurgood Marshall is reported to have said. "They mean don't go.") Eight years after President Truman proposed the integration of the military, less than two years since *Brown v. Board of Education* overruled "separate but equal," Welty's determined avoidance of social reality assumed a social significance of its own. *The Ponder Heart* turned her into Mississippi's favorite daughter—the besieged white public was only too happy to see itself in her adorable eccentrics—which proved to be excellent timing, because she now came home for good.

She hadn't any choice. Her mother had undergone cataract surgery in 1955, and she did not recover well, becoming frailer and ever more demanding until her death in 1966. Both of Welty's brothers died in the same span of time. Only many years later, in *One Writer's Beginnings,* did Welty acknowledge the heartfelt sacrifice and acute frustration of this period, her frantic scribblings at the steering wheel as she raced between hospitals. She began a novel about a big family reunion but hadn't the concentration to carry it through. After a lifetime of refusing to consider teaching—a profession too closely associated either with her mother or with old maids—such work became a solace. She began to lecture on writing whenever she was asked, and she was asked increasingly often. ("I'm so well behaved" is how she modestly explained her new campus popularity. "I'm always on time, and I don't get drunk or hole up in a motel with my lover.")

These college lectures and the needling questions Welty had begun to be asked about her lack of involvement in civil rights gave rise to her best-known essays—"Place in Fiction" (1956) and "Must the Novelist Crusade?" (1965)—defending her position as a nonpolitical writer in highly political times. There is not much to dispute in her argument for the writer's need to be well rooted in his or her native soil (citations: Proust, Jane Austen) or in her point about the damaging results that political crusading may have on fiction (standard whipping boy: John Steinbeck). But Welty comes very close to expressing disdain for any exercise of judgment, or any expression of anger ("even to deplore, yelling is out of place"), and this demand for restraint does seem tied, unavoidably, to her very political perception that "we in the South are a hated people these days." Better than to judge or to get angry, she claims, is to do as she does—to affirm, to write only with love. A noble dictum, but it is the opposite of what Welty actually did in the one extraordinary work that she produced in all these years, a story that turns her rationalizations upside down and shows the kind of brave and honest writer she could still be.

On the night of June 11, 1963, Medgar Evers was shot in Jackson. Upon hearing the news, Welty went to her desk and wrote a story in a single sitting. Perhaps it was just that this kind of rapid-fire response always prompted her best work, or else it was the act of opening her eyes to the world again that made the difference. But the story she wrote—"Where Is the Voice Coming From?"—is ablaze with the reconnection of words and purpose, as though the current of power in her early stories had never been diverted. A dramatic monologue, it takes place entirely within the mind of Evers's killer—a fictional figure, for the real one had not yet been arrested—and is foul and vicious and pathetic and full of lacerating hate. The story begins with a couple watching television:

> I says to my wife, "You can reach and turn it off. You don't have to set and look at a black nigger face no longer than you want to, or listen to what you don't want to hear. It's still a free country."
>
> I reckon that's how I give myself the idea.

The story appeared in *The New Yorker* two weeks after it was written. Details had to be altered for legal reasons, because some of the author's inventions were so close to the facts discovered in the few days after the killer was caught. But Welty had mistaken one significant detail about the killer, as she was nearly bound to do: his class. Although the white-trash voice of the piece was brilliantly sustained and chilling, the real killer was from one of the best Delta families. Yet as Welty had to point out herself, in defense of her intuition and the story's validity, the basic psychology of a murderous bigot is the same in every class, and this is it:

> There was one way left, for me to be ahead of you and stay ahead of you, by Dad, and I just taken it. Now I'm alive and you ain't. We ain't never now, never going to be equals and you know why? One of us is dead.

It was only in later interviews, explaining how she'd come to write this story, that she referred to "that world of hate that I felt I had grown up with." (It took three trials and thirty years to convict Evers's actual killer.)

Elsewhere, rather like Uncle Daniel Ponder, she increasingly insisted that she had been "brought up in a world of love." Her loyalty to a past and now often despised way of life was naturally intensified by a loyalty to the family she'd lost, but also, it seems, by her need to justify her years of sacrifice to them. (Compare Faulkner on the subject of the artist's sacrifice: "If a writer has to rob his mother, he will not hesitate; the 'Ode on a Grecian Urn' is worth any number of old ladies.") Welty produced just two short stories in the eleven years between her mother's surgery and her mother's death. Then she released a small torrent of work, all on the subject of the gloriously intractable bonds of family.

The Optimist's Daughter (1972) and *One Writer's Beginnings* (1984)—an autobiographical novel and a romanticized memoir—offer dual tributes to Welty's parents' perfect marriage. "Between some two people," she writes in the latter, "every word is beautiful." Their relationship seems to have loomed ever larger in Welty's mental life, and these books suggest wistfully overlapping images of the author herself: as a child standing lonely and listening intently outside a closed circle of adult love, and as an old woman even lonelier, still straining after the voices in an awful silence. Welty's family portrait does not omit the difficulties of having been her mother's daughter, but only hints at the continual interference, the guilt, and what seems to have been a battering anger in the later years. Yet these slender, plainly written books are as morally simplified as her earlier fables: the element of fantasy remains in the nearly absolute division of evil and good, which tend to be categorized by degrees of gentility. Racial tensions are absent because no one in the author's field of vision, black or white, questions his or her given place; kindly employers and grateful servants alike are bound by ties of steadfast loyalty, without a trace of the ironies or brackish undercurrents of an early story like "A Worn Path." Perhaps for just such comforting reasons, these late books have been among Welty's most popular, and have contributed to her growing aura as a kind of Eleanor Roosevelt of literature.

The *Collected Stories* of 1980 conformed to this tidied image by way of serious bowdlerization: with Welty's consent, the word "nigger" was removed from a significant portion of the work. (The 1998 Library of America edition restores the original texts.) Readers of this popular volume, which is still in print, will find that Uncle Rondo's belligerent niece commandeers a girl of no particular color to haul her belongings to the p.o., that there are "helpers" of no particular color fishing at Moon Lake, and that the old-maid piano teacher was raped—was this really thought to be an improvement?—by a "crazy Negro." Sainted Mrs. Rainey, on the other hand, is allowed to retain her fond thoughts about an old "trustworthy nigger," presumably on the principle that an affectionate tone renders racial condescension harmless—which was perhaps not so far from what Welty's message had come to seem by that time, after all.

Less tightly controlled than *The Optimist's Daughter* or *One Writer's Beginnings,* less readable, and far more interesting is Welty's big novel, *Losing Battles*—the "family reunion" book that was brewing during all those difficult years and was finally published in 1970. A sprawling, unwieldy work, it appears to have been written in a state of war between the author and her subject—or, more precisely, between the author and what she tries to feel about her subject. Welty spoke of *Losing Battles* as "a novel of admiration for the human being who can cope with any condition, even ignorance, and keep a courage, a joy of life." But all that affirmation is hard to find in this imbalanced, highly disturbing book.

The setting is the northern hills of Mississippi, where the land is so poor it never supported a plantation culture or developed a black population—there are no black characters in this book—and which Welty chose in order to concentrate entirely on family life; as a geographical decision this was not so different from *Delta Wedding*'s transference to a time when nothing much had happened, and it ensures an even greater social isolation. Despite her deliberately restricted agenda, however, Welty keeps veering off into dangerous territories—and then veers as quickly back again. It takes hundreds of pages of rustic farce to bring the members of a poor-white country clan together before we learn, in a few asides, their various secrets. One-handed, religion-spouting Uncle Nathan, for example, once murdered a man and "let em hang a sawmill nigger for it"; he cut the hand off himself, as part of his penance. The revelation is quietly stunning, and seems to matter, yet it is given no weight and immediately disappears.

Altogether, this book seems a puzzle in which the most important pieces have been hidden, even from the author. And it is precisely these pieces, however baffling or obscure or ruinous to the scheme, that give the book its power. There is, for example, a three-page scene involving fruit that is more sexually bizarre than anything since Christina Rossetti's *Goblin Market:* the women of the clan surround an overproud young wife and knock her down, pin her flat, then ram slabs of watermelon into her face—

"the red hulk shoved down into her face, as big as a man's clayed shoe, swarming with seeds, warm with rain-thin juice"—and down her throat. "Ribbons of juice crawled on her neck and circled it, as hands robbed of sex spread her jaws open." Their purpose—"Come on, sisters, help feed her! Let's cram it down her little red lane!"—is to get her to admit that she shares their name, and so is just like them, one of the family:

> "Why, you're just in the bosom of your own family," somebody's voice cried softly as if in condolence. Melon and fingers together went into her mouth. "Just swallow," said the voice. "*Everybody's* got *something* they could cry about."

The scene is so violent that a reader is not sure the victim will survive. As it turns out, her dress is barely soiled, and all is soon forgotten.

Forgotten, too, are the lessons of Welty's last and finest old-maid schoolteacher, the woman who loses the battles of the title and dies having tried and failed to bring enlightenment to Mississippi. The tale of her death is horrible: old and feeble, she is tied to her bed by one of the family aunts and, worse, her pencil and paper are taken away. That is when she really wants to die. For a short while, she scribbles with a wet finger on her bedsheet, but then the sheet is ripped off to make her stop. Somehow she manages a final letter, which is sneaked out to one of her former students, a judge now nearly an old man himself; it contains an urgent warning about the one thing she has failed to take into account:

> Watch out for innocence. Could *you* be tempted by it . . . and conspire with the ignorant and the lawless and the foolish and even the wicked, *to hold your tongue?*

Buried in the core of this misshapen, determinedly bumptious comedy is an American tragedy of true distinction: a tragedy of hidden knowledge and willed innocence. But Welty was not in the business of writing tragedies, and if she spent much of her life ignoring her heroine's warning, that life was permitted to run a far happier course.

In a public tribute to Welty in 1969, Faulkner's great champion Malcolm Cowley spoke in praise of her already famous qualities: "Gentle, unruffled, unassuming, kind, she is an unusual figure. . . . 'Isn't she nice!' other writers always say of her. Her writing is nice, too . . . fastidious, scrupulous, marked by delicate discrimination." And this is the image of

Welty that has prevailed, firmly shutting out the old, indelicate discriminations, the fooly-racky-sacky deliriums, the world of hate she split wide open and exposed with expert hands. But clearly, this change has been her choice. A born outsider in a stifling, hypocritical, yet tantalizingly charming society, Welty discovered that she could write her way into acceptance; year by year, book by book, she came to be wholly embraced. Who would not find such acceptance irresistible? Who would not be glad and grateful? In terms of her art, however, it was a hard bargain. In the course of a long and proud career, an intrepid explorer turned herself into a perfect lady—a nearly Petrified Woman—with eyes averted and mouth set in a smile, who yet from time to time let out a bloodcurdling cry, as though she, too, could not forget what she had lost.

The Rage of Aphrodite

Marina Tsvetaeva

Before the Revolution, Marina Tsvetaeva refused to accept money for the publication of her poetry. She was not above accepting the tribute of a small gift, however, and it was as a guest of the publisher of a St. Petersburg literary journal that, at the end of 1915, she travelled from her home in Moscow to a New Year's party and private reading in the imperial capital, where, as a blizzard raged, "*all* Petersburg read, and *one* Moscow," she reported proudly. At twenty-three, Tsvetaeva was already the author of a volume of poetry that had been acclaimed for its apparent spontaneity and unabashed intimacy—it was her first book, which she'd published herself, at eighteen—and of a second book, which had been summarily dismissed. Her work was subject, understandably, to emotional awkwardness and to flourishes of overstatement; critics, noting her age and her command of a range of poetic forms, expressed the hope that she would outgrow her weaknesses. But Marina Tsvetaeva outgrew nothing. Instead, she cultivated her weaknesses as she did her strengths. She clung to emotional extremes, she intensified her use of hyperbole and her tone of outcry, and, in the process, she developed the most peculiarly excitable and brilliant and perhaps the most individual style in twentieth-century Russian poetry.

The supreme Russian poets whose works bridge the Revolution—Anna Akhmatova, Boris Pasternak, Osip Mandelstam, and Tsvetaeva—were born within three years of one another, between 1889 and 1892, as though history had determined to provide an immortal record of the devastation to come. Only Akhmatova and Pasternak survived into old age. Mandelstam was killed by his government in 1938, and Tsvetaeva died—there are those who would say she, too, was killed—in 1941. In recent years, the poetry and essays and letters of these great poet-martyrs have been published in enormous editions and subjected to intensive study in Russia and, increasingly, in Europe and the United States. In this country, Tsvetaeva remains the least known of the illustrious quartet (a grouping acknowledged by Akhmatova in the late poem "The Four of Us"), and for good reason. There exists no commonly accepted English translation of even a fraction of her more than two thousand poems, and the majority of English versions that can be found in anthologies or in various slim selections do little to account for her veneration among Russian readers. Tsvetaeva's poetry not only is syntactically complex but is uniquely fixed in the sound patterns of the highly inflected Russian language: extraordinarily compressed and involute, with an explosive density of internal reference. *"S novym godom—svetom—kraem—krovom!"* she begins the poem "Novogodnee" ("New Year's"), an elegy to Rilke, and Joseph Brodsky's deliberately rough working translation gives us "Happy New Year—World/Light—Edge/Realm—Haven!" This is a poetry of driven rhythms and steep intonation, of high musicality. "They absolutely must be read aloud, otherwise they'll fall flat," Tsvetaeva warned of one group of lyrics, and the injunction may be taken broadly.

Simon Karlinsky, whose groundbreaking 1966 study of her life and work ushered Tsvetaeva into the English-speaking world, dismissed even the possibility of translating certain of her effects. In what seems an ironic corroboration, Brodsky has written persuasively, in an essay available in English, on the originality and power of Tsvetaeva's Russian versification in the Rilke elegy, a work for which no remotely satisfying English translation exists, or perhaps can exist. (After more than six pages of analysis of the single "overloaded" opening line, Brodsky notes, "There are 193 more lines like this in 'Novogodnee.' ") Bilingual editions of her White Guard poems, translated by Robin Kemball as *The Demesne of the Swans,* and of her last collection, *After Russia,* translated by Michael M. Naydan with Slava

Yastremski, can serve as tools for those who might work their way into the Russian or simply try to get a sense of it (this reader included), but they provide few convincing, freestanding English poems. (A mere glance at the juxtaposed columns of these editions reveals how Tsvetaeva's compounded blocks of words—often just one or two words to a line—are demonumentalized in translation, shattered into the requisite English shower of verbs and prepositions.) The fact is that Tsvetaeva's language is difficult even for many Russians. Rilke, who in his youth had tried his hand at composing poems in Russian, but whose command of the language had faded by the time he corresponded with Tsvetaeva, had to admit to her that while he prized her as a poet—through her letters (written in German) and through his sense of an essential rhapsodic affinity—he couldn't actually grasp her poetry at all.

None of this is to say that Tsvetaeva's work is "difficult" in the sense of being dry or in any way abstruse. On the contrary, she is the most rivetingly direct and dramatic of poets, feverish with the need to communicate. And her most abiding subject is the torment of passionate, sexual love, a love not meekly awaited or gratefully accepted but demanded as a right and held out as a goal or—in poems addressed to men she briefly worshipped or desired—as a dare. Next to the keening violins of women's traditional love poetry, brought to perfection in Russia by Akhmatova, Tsvetaeva's instrumentation is for brass: bold, charged, indignant. The brazen tone that once caused her work to be termed "hysterical" and condemned for "morbid eroticism" by offended contemporaries has now secured her entrance into the ranks of feminist heroines, fully sexed and unashamed.

Along with the appearance of this "new" Tsvetaeva in academic studies—Barbara Heldt's *Terrible Perfection: Women and Russian Literature* is a particularly able and insightful example—there have been several full-length biographies of Tsvetaeva published in English since Karlinsky's introductory effort. What has brought on such a show of attention? Even the marvellously sympathetic accounts of her life and work by Viktoria Schweitzer (*Tsvetaeva,* with translations of the poems by Peter Norman) and Jane A. Taubman (*A Life Through Poetry: Marina Tsvetaeva's Lyric Diary*) convey the poet's themes and preoccupations and a sense of her psychology (albeit an entirely contrasting sense) but not—by their own admission—the formal brilliance or originality of the poetry itself. Two collections of Tsvetaeva's irresistibly vivid essays have appeared in English since the 1980s—titled "Art in the Light of Conscience" and "A Captive Spirit"—but these remain little known. Interest was piqued by the publication of her correspondence with Rilke and Pasternak, in German in 1983 and in English two years later, after a ban she had imposed on the publication of Rilke's letters had expired ("when Rilke's letters will be simply Rilke letters—not to me—to everybody"). But is it possible to embrace a poet simply as a stirring figure without embracing, or even being able to read, the poetry? In our insatiable age of biography, it may be so.

The most proudly Russian of poets came from a line of poor village priests of Vladimir province on her father's side—he told of having had no shoes of his own until he was twelve years old—and from Polish nobility on the side of the young woman who had knowingly renounced all happiness upon her marriage to him. Maria Alexandrovna Meyn was twenty-one years old, a gifted pianist (a student of a student of Anton Rubinstein's) whose father had forbidden her to go on the stage, a blazing romantic whose love for a married man had threatened her family with scandal—the man sought a divorce, but Maria's father would not hear of such an association—when, in 1891, she gave up her battles and wearily agreed to marry an aging widower with two young children. Ivan Validimirovich Tsvetaev, then in his middle forties, had studied hard and risen to the position of professor of Roman literature at Moscow University, and had embarked on his life's task of building a museum to house the university's collection of replicas of ancient sculpture; a kindly man, entirely deaf to music, he does not seem to have had a clue as to the nature of his wife's feelings. But her daughters knew everything: Marina, born in 1892, and Anastasya, born two years later, received the full benefit and bore the full brunt of their mother's subjugated passions for music and poetry—"Mother gave us drink from the opened vein of Lyricism," her elder daughter wrote years later—and for heroes. ("When Rubinstein shook her hand, she wouldn't remove her glove for two days," Marina also recalled.) It was of course Marina, with perfect pitch and a talent for language obvious by the time she was four, who became the special object of her mother's ambitions. Her early childhood years, memorialized in an essay titled "Mother and Music," were a martyrdom to the enormous black piano at which she first got to know the reflection of her own face and at which she suffered every day from her inability to meet her mother's inexorable demands. ("My mother who demanded from me—herself!")

At the age of six she was taken to a performance of *Eugene Onegin* at her music school and, she believed, her emotional fate was sealed. The example of Tatyana's headlong declaration of love to Onegin prompted her to write a "Tatyana" letter to her stepbrother's tutor (it was returned with grammatical errors corrected), and doomed her ever afterward to be the first to make the imploringly seductive gesture and the last to expect to see it returned: "That first love scene of mine foreordained all the ones that followed, all the passion in me for unhappy, non-reciprocal, impossible love." But, thinking further, she came to believe that her lamentable fate had been sealed much earlier, even before her birth, by her mother's decision to follow the example of the very same Tatyana—the only other and even worse choice of heroine, she noted, would have been Anna Karenina—in her final resignation to a deadening but dignified married life with a man she did not love, a decision that ultimately produced Marina Tsvetaeva herself: "If there had been no Pushkin's Tatyana, I would not have come into existence. For women read poets *that way* and not otherwise." Thus the most proudly Russian poet's true line of descent was from Russian poetry.

When, in 1902, Maria Alexandrovna was diagnosed with tuberculosis, she took her girls abroad for a period of three years. Living in a Russian boardinghouse near Genoa, Marina fell under the sway of a group of exiled anarchists; aged ten, she wrote poems of revolutionary fervor that she still recited thirty years later. Sent on to boarding schools in Switzerland and Germany, she renounced atheism for an equally short-lived Catholicism and developed the long-lasting, rapturously pro-German sentiments that filled her poetry and scandalized Russian audiences during the First World War. Tsvetaeva had no ideology save romance and provocation, and there was nothing of her childhood she ever willingly gave up. Mother and daughters returned to Russia in 1905, settling in the Crimea for the climate; Maria Alexandrovna could hardly walk by then, but when she went home to die, in June 1906, she insisted on entering the house on her own and, for the last time, made her way to the piano. Marina Tsvetaeva would not reveal what it was her mother played then—that was to remain a secret between them—but she recorded her mother's final words, "I shall miss only the music and the sun." And she added that there was one important way in which Maria Alexandrovna did not die: "Her tormented soul lives on in us, but we reveal what she concealed. Her rebellion, her madness, her longing have grown in us to the level of a scream."

The tyranny of music in Tsvetaeva's life now came to an end, except as she learned to impose its rigors on her poetry. (Her frustrated translators would doubtess sympathize with a rebuke she received from the poet Konstantin Balmont: "You demand from poetry what only music can give!")

Her formal education also soon came to an end, as she was expelled from or left one school after another; she was never a dedicated student, and her academic career had turned into a series of crushes on older girls whose beauty she envied and, by a process of poetic inversion, adored. She covered the image of Christ on the icon in her room with a picture of Napoleon (although she worshipped Rostand almost as much), and at sixteen she spent a summer alone in Paris. At seventeen, back in Moscow, she bobbed her hair and took up smoking. She published her first book of poems at eighteen, "as a substitute," she later explained, "for a letter to a man with whom I was denied the possibility of communicating any other way." There is some suggestion that in this same year, 1910, she contemplated suicide—this information derives from her sister Anastasya's memoirs, written decades afterward in a spirit of retaliation and far from certainly reliable—by shooting herself during a performance by Sarah Bernhardt in Rostand's *L'Aiglon* on tour in Moscow (the last performance, presumably, as she had purchased tickets to every one). "The fault lies with books," Tsvetaeva wrote, "and with my deep mistrust of real, everyday life."

It seems understandable that those who knew her have often reported that there was no division between Tsvetaeva the poet and Tsvetaeva the woman. Each was emotionally outsized, unflinching, continually and insistently risking a spill into foolishness or disaster instead of the ecstatic achievement she sought. She drove most people around her half mad. For Tsvetaeva's greatest passion was always for her poetry, and for the extremes of experience that made the writing possible. In a 1928 poem entitled "A School for Verse" (or, alternatively, "Conversation with Genius"), the offshoot of a private course in poetics she administered to a charming and highly literate young mountain climber, Tsvetaeva protests—and answers for—the almost intolerable burden of her vocation:

> "It's torture!"—"Endure!"
> "My throat's a mown meadow!"
> "Wheeze:
> That too is a sound!"
>
> "It's a business for lions,
> Not women." "It's child's play;
> Even disemboweled—
> Orpheus sang!"

"And so to the grave?"
"And from the grave, too."
"I can't sing!"
"Sing of that!"

(Translated by Jane A. Taubman)

Tending her genius, abetted by history, Marina Tsvetaeva lived the life her poems required.

She was, first of all, a poet of Moscow, so sure and exuberant in the possession of her native city that as a young woman she habitually bestowed it, embodied in her poems, on whomever she loved. Most of all she loved the supremely accomplished, snobbish, "foreign" poets of St. Petersburg—Aleksandr Blok and Akhmatova and Mandelstam—who led the literary field, and for whom, in their yearning toward European culture, her Moscow was no more than a provincial backwater. The party in St. Petersburg that Tsvetaeva attended on New Year's of 1916 was her chance to prove not only her own worth but the possibility that a poet might come out of Moscow and, still more astonishing, choose to go back. Despite her suspicion that to some of the more refined local ears she sounded like "a Moscow coachman," she was aware of having a great success and of "elevating the name of Moscow to the level of the name of Akhmatova."

Anna Akhmatova was the dark lady of literary Petersburg—the Garbo of Russian poetry—whose melancholy glamour and poems of wounded love had won her a huge cult. Although Akhmatova was not present at the reading, she remained for Tsvetaeva the measure of her performance and of how she defined herself. Akhmatova was then only twenty-six to Tsvetaeva's twenty-three, but their personal hierarchy was fixed—less, perhaps, by any disproportion between their gifts or their reputations than by the lifelong propensity of Akhmatova to be adored and of Tsvetaeva to do the adoring. (Although Akhmatova was a legendary beauty and Tsvetaeva declared herself "shamefaced at my own plainness," the photographs reveal no such extremes; the impression each made was more a matter of affect.) Later that year, after her trip to St. Petersburg, Tsvetaeva wrote a series of eleven poems to Akhmatova—addressing her, "O, Muse of lamentation, most beautiful of muses!"—in which she brashly but formally presented her with the city of Moscow "and my own heart in the bargain," dispatching both to an apparently indifferent recipient. There was no response; or, rather, there was a response that took twenty-four years to appear. For the moment, Tsvetaeva comforted herself with a story, told her by Osip Mandelstam, that Akhmatova carried the poems around with her everywhere, and that, upon hearing a description of the "remarkable" intruder on her territory, Akhmatova had demanded, impatiently, "But can one fall in love with her?"

In fact, Mandelstam had fallen in love with Tsvetaeva, apparently won over by the stubborn loyalty she displayed in reciting poems in praise of Germany while the Great War raged and while the Dutch name of Sankt Peterburg itself had been Russianized to Petrograd. The most exquisitely Petersburgian of poets—which is to say, the most classically pure and cerebral—Mandelstam was also, at twenty-five, the most tender, and he was soon travelling back and forth between his city and Tsvetaeva's so often that an acquaintance asked whether he worked for the railroad. During the late winter and spring of 1916, Tsvetaeva led her "foreign guest" on a path of discovery through her ancient, long-deposed capital, and wrote for him poem after poem filled with the heavy allure of gold domes, dark chapels, and pealing bells.

She began one poem, quite simply:

> Take from my hands, my strange,
> my beautiful brother,
> This city not made by hands

And, accepting the gift, the great classicist replied:

> In the walls of the Acropolis sorrow
> consumed me,
> For a Russian name and Russian beauty.

Mandelstam, for his part, was writing for the first time of Russia and, with far greater hesitancy, of love. (His most famous poem up until then begins, "I am given a body—What am I to do with it?") Tsvetaeva saw from the start that, while the poems they wrote for each other might last an eternity, the fretful young poet himself wouldn't alight for long. She raised no objections, but wrote of her "divine boy" at times with a teasing and almost Hellenic sweetness, as though mimicking his own calm tradition:

Where does this tenderness come from?
These are not the—first curls I
have stroked slowly—and lips I
have known are—darker than yours

as stars rise often and go out again
(where does this tenderness come from?)
so many eyes have risen and died out
in front of these eyes of mine.

and yet no such song have
I heard in the darkness of night before,
(where does this tenderness come from?):
here, on the ribs of the singer.

Where does this tenderness come from?
And what shall I do with it, young
sly singer, just passing by?
Your lashes are—longer than anyone's.

(Translated by Elaine Feinstein)

When Mandelstam indeed ran off for good that summer of 1916, fleeing a shared country idyll with hardly a warning, Tsvetaeva's liberating gifts to his poetry—what Nadezhda Mandelstam acknowledged years later as a new breadth of feeling and style—stayed with him. This was what she had promised him, in order that, as she concluded one poem, "you won't repent that you loved me."

Free spirit though she was, Tsvetaeva was not unconcerned with repentance, for she was by this time a married woman and the mother of a little girl. She had met her husband, Sergei Efron, at the Crimean beach resort of Koktebel in 1911, when he was seventeen, a tragic-eyed boy making a shaky recovery from tuberculosis. An aspiring writer himself, Efron was duly impressed with this eighteen-year-old paragon who had already published a book of poetry. The two spent the summer together, writing poems and stories about their childhoods—neither having yet acquired much more of a subject—and were quietly married in Moscow in late January 1912. Ariadna Efron, known as Alya, was born in early September.

There were no new poems then for a while, until Tsvetaeva began writing to her year-old daughter. As a source of inspiration, however, Ariadna quickly yielded to the poet Sofia Parnok, with whom Tsvetaeva conducted an open lesbian affair in 1915 while writing lines like "my beloved sisters, /

we shall certainly find ourselves in Hell!" (and while her anguished husband moved in with his sister). Then Parnok yielded, in her turn, to Mandelstam. "My verses are my diary," Tsvetaeva proclaimed. "My poetry is a poetry of proper names." Sergei Efron, too, inspired his share of poems: works of gentle regard, notably sexless, concerned with nobility and innocence.

In just a few years, Tsvetaeva would write a play about a young woman who awakens to the awful knowledge that she doesn't love her husband, and decides to run off with a wise, older man, who turns out to be a vision and simply disappears. Tsvetaeva, convinced throughout her life that her "Seryozha" could not get along without her, veered wide of her marriage but did not, in fact, run off. (It was left to Akhmatova to pronounce, on the basis of her own experience, that the "institution of divorce was the best thing mankind ever invented.") All it took was for the latest vision to disappear—they always disappeared—and she returned home. Whether as the cause or as the result of one of her chronic marital reconciliations, Tsvetaeva gave birth to a second daughter, Irina Efron, on April 13, 1917, three days before Vladimir Lenin disembarked at Petrograd's Finland Station.

Tsvetaeva had written her first antirevolutionary poems within a week of the February uprising in Petrograd. The phone lines between the two cities had been cut, and lumbering along in the late stages of pregnancy, she went out among the frenzied crowds filling the Moscow streets as the news spread and saw there only ashen, indistinguishable figures: "They have no faces, they have no names / They have no songs!" On the day of the czar's abdication, she recorded a prayer that Moscow might lie down "to eternal sleep!" For Tsvetaeva, such thoughts represented not a political position, in any standard sense, so much as an emotional and even an aesthetic one. By the fall of 1917, she was inveighing against the "Byzantine perfidy" of Nicholas II, and she came to blame no less a figure than Peter the Great ("Sire of the Soviets . . . Sire of the rubble—it's *your* doing— / These monasteries in flames!") for setting the course of squandering and high-handed carelessness that had led directly to the horrors of October.

Tsvetaeva actually more or less missed the October coup—at least as it was carried out in Moscow. Efron was serving as an instructor in the infantry reserve, his fragile health deteriorating under the strenuous work.

Tsvetaeva had travelled to the Crimea for a respite from domestic responsibilities and in the hope of finding a tranquil place where the family could pass the winter. But her plans came to nothing, and after a few weeks she boarded a train back to Moscow, ignorant of the events that had occurred in her absence. The first newspapers she saw made the situation clear, detailing the Bolshevik takeover and the street fighting, and suggesting casualties in the thousands; the regiment still holding out around the Kremlin was Sergei's. Her journey home was a slow agony, marked by continual delays and dead stops, by passengers with vying rumors of chaos and reprisal. Fearing the worst, she wrote a long and adoring letter to her husband: "If God performs this miracle and leaves you alive, I will follow you like a dog."

In Moscow, Efron hung on until the final surrender. The next evening, husband and wife boarded a train back south, she accompanying him as far as she could on his way to join the White Army along the Don. Then Tsvetaeva set out yet again for Moscow, in order to fetch the children back to the comparative haven of the White-controlled Crimea, near their father. But this time, when she reached the stricken city, all escape routes closed behind her.

A twenty-five-year-old woman not weak but unworldly almost by vocation, who neither professed nor displayed any skills beyond the writing of verse, whose parents were dead and whose husband might well have been, Tsvetaeva was lacking, moreover, in any instinct for cunning or self-preservation, or even for what might be called mere getting along; and she now had to get along not only for herself but for two small children in a devastated world. In 1918, the year her family house was dismantled for firewood, she wrote several plays for the Moscow Art Theatre, all unproduced, and secured translating work for Meyerhold's theatre company, but she was soon fired for asking too many questions that showed her to be, in Meyerhold's words, "hostile to all that is sanctified by the idea of the Great October." After this she managed by selling her dwindling possessions, by pocketing food from the tables of friends, and by reciting, for a small fee, at some of the poetry readings that had long been a part of Russia's richly vocal poetic tradition but had taken on a new urgency in those early Soviet years, when publishing nearly ceased for lack of paper and ink.

Still, Tsvetaeva was fully capable of refusing her fee for a reading—as a protest against the fee's smallness and against the new society's failure to value a poet's work. And the poems that she did read in public, quite fearlessly, were often the rhetorically overcharged paeans to the White Army which she was composing in the name of her distant and heroic husband: "Gallant White Legions! White stars that, steep, inde- / Faceable, span the skies!" At the same time, more privately, she was turning out impassioned poetic tributes to various men on so frequent a basis that a single poem might have to bear up under several sequential dedications. Nor were her loyalties irreconcilable with a poem she called simply "Bolshevik," its subject a stalwartly unseducible eighteen-year-old Red soldier—his shoulders stretched "From the Ilmen to the Caspian waters"—just back from Crimean battlefields. For Tsvetaeva, the literal embrace of the men of both sides was—at least, politically—not an evasion of principle but the embrace of a principle beyond politics:

> This man was White now he's become Red.
> Blood has reddened him.
> This one was Red now he's become White.
> Death has whitened him.
>
> *(Translated by Elaine Feinstein)*

By 1919 the squalor into which Tsvetaeva and her children had plummeted was notorious. Her friend Prince Volkonsky, former director of the imperial theatres, wrote in his memoirs that a thief had broken into Tsvetaeva's apartment and been so horror-struck that he'd offered her money. She herself described her apartment's state of "cold, puddles, sawdust, buckets, pitchers, rags—everywhere children's dresses and shirts." Each day, she went on a round of institutions to collect enough food for her girls, then aged seven and two, and that autumn, faced with the limits of her endurance and the terrors of winter, she put the children in an orphanage outside Moscow and moved in with friends. Alya, the elder and favored daughter—her mother's trained companion, a small fount of memorized verse—soon became ill and had to come home to be nursed. While she was recovering, in February 1920, Tsvetaeva received notice that little Irina had died of starvation.

There followed two months of very few words, after which Tsvetaeva wrote a single poem about her lost child, insisting that, as a mother, she had held on "fiercely, as best I could!" She then buried her grief in a new love affair—several critics have registered dismay and even horror at the rapidity of her recovery—which yielded a new cycle of poems, twenty-seven in all. Certainly life resumed, even a little easier than before. Admirers saw to it now that she was supplied with a ration card, and she could afford once again to be generous. (When a young poet named Mindlin moved in with her for a time, her newfound domestic abilities caused a friend to protest, "Why wash *his* shirts? He's a terrible poet.") And then, in July 1921, several months after the White Army's final defeat and nearly a year since she had heard any news of Sergei Efron, a letter finally arrived. She started a new poem at once—"Alive and Well!"—even as she made plans for her departure to meet her husband in Berlin, feeling "like a bull, / Under the axe head / Of happiness."

The evacuation of the White Army from Russia had left Efron in Constantinople, and he had gone on to settle in Prague. By the time he made his way to Berlin to meet his wife after a separation of four and a half years—"I need nothing, except the fact that you are alive," he had written her—she was too caught up in an affair with her new Berlin publisher to express the sentiments that might reasonably have been expected. Her diary does not even record their reunion. Efron quickly retreated to Prague, where he had been given a stipend for study at the university (as were all displaced Russian students who applied). Almost as quickly, Tsvetaeva broke with her publisher (soon to be identified in her correspondence as "the black velvet nonentity") and, packing up her life yet again, doggedly set out after the man she had driven away.

Russian writers in Czechoslovakia were also awarded stipends, and although Tsvetaeva lived what Efron called the "life of a camel" there for three years, in a muddy outlying village, she was enormously productive, turning out a quantity of lyric and longer narrative poems and the first play of a verse trilogy she planned to call "The Rage of Aphrodite." Her dramatic heroines were Ariadne, Phaedra, and Helen, and her subject was the compulsions and costs, for women, of erotic love. (Only two plays were completed. Vladimir Nabokov, who met Tsvetaeva in Prague and became an admirer of her work after her death, wrote that her *Phaedra* caused him "astonishment and a severe headache.")

She knew her subject. In 1923, Tsvetaeva suffered vigorously, almost professionally, first over a young critic in Berlin whom she knew through an admiring review of her poetry and then through his letters, and whom her intensity simply frightened away. ("I cannot tolerate the slightest turning of the head away from me," she warned him. "I HURT, do you understand? I am a person skinned alive, while all the rest of you have armor.") That fall, she moved on to a full-blown affair with a friend of Sergei's named Konstantin Rodzevich, which turned out to provide a high point of private anguish and poetic accomplishment. The most unknowing and unconcerned of muses, Rodzevich also quickly found Tsvetaeva's intensity unbearable, and he is spied in her poetry chiefly in the posture of escape—in the long psychological narrative called "Poem of the End," about their final walk together through the streets of Prague, and in the poem that has become Tsvetaeva's most popular and frequently anthologized work, "An Attempt at Jealousy," written after a chance encounter with Rodzevich and his suddenly acquired fiancée in 1924:

How's your life with the other one—
Easier?—A stroke of the oar!—
Like the coastline
Does it take long for the memory to recede

Of me, a floating island
(In the sky—not on the waters!)
Souls, souls! you should be sisters,
Never lovers—you!

How's your life with an *ordinary*
Woman? *Without* deities?
Now that you've dethroned your Queen
(Having stepped down yourself).

How's your life—do you fuss—
Do you shiver? How do you feel when you get up?
How do you deal with the tax
Of deathless vulgarity, poor man?

"Convulsions and irregular heartbeat—
I've had enough! I'll rent my own place."
How's your life with anyone—
My own chosen one!

More compatible and more palatable
The food? If it palls—don't complain . . .
How's your life with an imitation—
You who have tramped upon Sinai!

How's your life with a local
Stranger? Point-blank—do you love her?

Or does shame, like Zeus' reins,
Lash at your forehead?

How's your life—your health—
How've you been? Are you managing to sing?
How do you deal with the ulcer
Of deathless conscience, poor man?

How's your life with goods
From the market? Is the price steep?
After Carrara marble
How's your life with plaster

Dust? (From a solid block was hewn
A god—and smashed to bits!)
How's your life with one of a hundred-thousand—
You, who have known Lilith!

Are you sated with the newest thing
From the market? Having cooled to magic,
How's your life with an earthly
Woman, without a sixth

Sense? Well, let's hear it: are you happy?
No? In a shallow bottomless pit—
How's your life, my darling? Harder than,
Just like, mine with another man?

(Translated by Barbara Heldt)

Tsvetaeva's long seizure of passion for Rodzevich nearly tore Efron apart, and he began now to ponder going back to Russia—as "a wounded animal crawls back into its lair," he wrote to a friend. He saw his wife clearly in her "hurricanes" of self-deception, yet he could not bring himself to leave: "I am at one and the same time both a lifebelt and a millstone round her neck. It is impossible to free her from the millstone without tearing away the only straw she still has to clutch. My life is utter torment."

Tsvetaeva was six months pregnant when she wrote "An Attempt at Jealousy." She was ecstatic when, in February 1925, she gave birth to a son: she had dreamed of a strong, heroic boy child at least since Irina's death. Georgi Efron was lovingly nicknamed Mur—after a clever tomcat out of E. T. A. Hoffmann—and he became the pride of her life, the only force in it equal to her poetry. There was, of course, much gossip in Prague circles about Rodzevich and the child's paternity. In fact, Efron had vetoed Tsvetaeva's desire to name her son for yet another man—one who clearly was not the child's father. "He was Boris for nine months inside me and ten days outside," she wrote in explanation to Boris Pasternak, who was becoming the pivotal figure in her intellectual and emotional experience even though, by her contrivance, they were almost never to meet. She had finally learned to beware the flesh, the loss inherent in consummation. During the months before childbirth, looking back on Rodzevich and all the others, Tsvetaeva wrote, "They have *always* parted with me that way, except B.P., with whom my meeting—and *consequently*, my parting—is still ahead."

"My dear, golden, incomparable poet," Pasternak had saluted her in his first letter, sent from Moscow to Berlin in June 1922. He had just read her most recent collection. Rushing out to find and read his poems in her turn, Tsvetaeva responded with a reciprocally effusive letter, and also with her first critical essay, published in a Russian expatriate journal: "This is not a review. It is an attempt at release, so I don't choke. I've discovered the only contemporary for whom my lungs are inadequate."

Tsvetaeva and Pasternak were natural counterparts. Passionate Muscovites, both had grown up steeped in German culture; each worshipped Rilke as the Orpheus of the age. Both had received strict musical educations—Pasternak's mother was also a pianist—and the manipulation of pure sound into meaning marks their work in similar ways. And, perhaps most important, both were consumed with the transcendent aspect of the poet's calling, and found reality a near impossible premise for living.

Pasternak, recently married, came to Berlin with his wife in the summer of 1922; Tsvetaeva left for Prague a few days before his arrival. "My favorite kind of relationship is otherworldly: to see someone in a dream. And the second is correspondence," she wrote him. But when, the following March, he gave up on his attempt at exile and decided to return to Moscow, she was filled with anguish: the news had come too late for her to get to Berlin to say good-bye. In the days before Pasternak's departure and in the weeks afterward, Tsvetaeva wrote a lyric cycle to him of ten poems—poems filled with a hard, bruising force of desire rarely associated with the love poetry of women:

> So suffer me through and survive me! I am everywhere:
> I am dawns and ores, bread and breath;
> I am and I shall be and I shall obtain
> Your lips the way God will obtain your soul.

This is a fragment (translated by Simon Karlinsky) of a longer poem, as is the following (translated by Peter Norman), which was written two days later:

> Patiently, as stone is crushed,
> Patiently, as one waits for death,
> Patiently, as news ripens,
> Patiently, as revenge is cherished—
>
> I shall wait for you.

Tsvetaeva waited for Pasternak, in her fashion, throughout her Prague years, and in late 1925 she moved to Paris with her family and waited there. They had gone in the hope of finding a larger audience for her work in the city that had taken Berlin's place as the center of Russian émigré culture, and Tsvetaeva was soon publishing in all the relevant journals. She drew a tremendous crowd—as though for a performance by Chaliapin, Efron noted—at a public reading she gave in February 1926. Some of the excitement surely had to do with the political spirit of the White Guard poems that she often read, and also with the dramatic power of her voice, which was, by contemporary testimony, overwhelming.

But, success or no, the family remained stuck in poverty. Efron, as usual, had no work, and Tsvetaeva wrote poetry less and less frequently, turning instead to essays, which paid better. In the spring of 1926, she succeeded in estranging herself from almost the entire Paris *russe* community with an angry polemic called "The Poet on the Critic," in which she attacked by name those influential arbiters who judged poets according to their politics. She argued that certain poets in the Soviet Union were writing not only the best Russian poetry of the day but poetry that did full honor to the sacred tradition that the Soviets were presumed to defile. Tsvetaeva saw her task as the defense of genius: the genius of the daredevil Bolshevik spokesman Vladimir Mayakovsky and, above all, of Pasternak.

"Four evenings in a row I have thrust into my coat pocket a fragment of a haze-moist, smoke-dim Prague night, now with a bridge in the distance, now with you there, before my very eyes," Pasternak wrote her from Moscow that March, after coming across a privately circulating, unpunctuated typescript of her "Poem of the End." The faltering of his marriage doubtless made him especially susceptible to Tsvetaeva's epic of love's exhaustion, and to its author. Within a month, he announced his willingness to leave his wife, child, and country: "Answer me as you have never answered anyone before, as you would answer your own self. *Shall I join you now or within a year?*" A year was the limit he could bear, he told her, and the time would, moreover, be useful for completing some vitally important work. Trembling on the brink, he warned, "If you don't stop me I will come now, and *only to you,* but empty-handed and with not so much as an inkling as to where to go from there or why."

It seems a kindness that she answered with a letter he found "chilling," informing him that a visit at the present time would not fit in with her plans to take her children to the seashore. In later years, she referred to his proposal as an averted "catastrophe" for their families. But, more than that, her poems to him—and his greatest value to her was in her poems to him—depended on longing, and therefore on absence. The poet already had what she wanted, and mere happiness would have made a poor exchange. Pasternak continued, briefly, to advance his suit but soon settled into a grateful "You sat me down to work," and promised to send her his latest poems. In the meantime, as proof of his devotion, he was making her the extraordinary gift of Orpheus himself; that is, of Rainer Maria Rilke.

Pasternak had only recently begun an exchange of letters with Rilke, who was living in isolation in Switzerland, and it was at Pasternak's request that Rilke now sent copies of his *Sonnets to Orpheus* and *Duino Elegies* to Tsvetaeva. She responded eagerly and sent him her own "Poems to Blok." By mid-May 1926, the two were in regular correspondence. But Pasternak's gift did not stand him in very good stead; her letters to him now became cruel, edged with contempt. To Rilke, the angelic poet, Tsvetaeva could write with loving discernment, "You alone have said something new to God." That she was soon confounding angel and man—seeking a body for that most disembodied of poetic voices—is ironic but perhaps inevitable; she took three years off her age (she was thirty-three) when she described herself to him, and assured him, too, that she weighed *"nothing."* Within months, she was writing to him, almost begging: "Rainer, I want to come to you. . . . I want to sleep beside you. . . . And nothing else. No, more: to sink my head into your left shoulder . . . nothing else. No, more: and even in the most profound sleep to know you are beside me." She proposed that they meet somewhere in the French Savoy.

Rilke, who had been as welcoming and nearly as hyperbolic up to this moment, now answered gently but evasively. He was concerned that he had come between her and Pasternak, and he was troubled by her demands of exclusivity. Recognizing the familiar signs of withdrawal, she justified her desires. "Love lives on exceptions, segregations, exclusiveness," she wrote, but she reassured him, too, that her words had been merely "a manner of speaking. A manner of loving." And, with a wretched honesty (if it is honesty), "Do you imagine that I believe in Savoy? Oh, yes, like yourself, as in the kingdom of heaven. . . . Dear one, don't be afraid, simply answer yes to every 'Give'—a beggar's comfort, innocent, without consequences. . . . The word, which for me already is the thing, is all I want."

Rilke was then fifty years old, and was aware that he was ill, although he didn't yet know that the illness was leukemia. He had written her seven letters and one major poem—"Oh the losses into the All, Marina, the stars that are falling!" And now the words stopped. She assumed that he was angry with her, that she had driven him away. Perhaps she had, because in these last months of his life, despite his ever more debilitating weakness, he did answer, through a secretary, the letters of other people. But not hers. She learned of his death just before New Year's (Rilke died on December 29) from some people who had come to invite her to a party. Her final letter to him, in the form of the great New Year's elegy—*"S novym godom—svetom—kraem—krovom!"*—was completed in February.

In their grief over Rilke and in honor of his memory, Tsvetaeva and Pasternak began to write each other again after months of bitter silence, but the results were predictably strained and fitful. Still, her unchanging belief in Pasternak's achievement came through in two luminous essays she wrote about him during the following years. "Mayakovsky acts *upon* us, Pasternak—*within* us," she said. "Pasternak isn't read by us: he takes place in us." But her path was diverging from his ever more widely. After 1930, when Mayakovsky committed suicide—a stunning rebuke to the system that had banked on his performance as the buoyant revolutionary—Pasternak became the officially lauded Soviet genius. Showered with privileges (and, eventually, a large dacha), the hero of one Stalinist writers' congress after another, he yet retained the sympathy of his noncomplying old friends—now largely outcasts and victims—because of the evident confusion and torment he felt at finding himself in such a position. (The greatest Russian translator of Shakespeare's tragedies, Pasternak played the Hamlet of the Revolution, much as Mayakovsky had been its Mercutio.)

In 1935 Pasternak was ordered to Paris as a guest of honor at the International Writers' Congress in Defense of Culture, an antifascist, pro-Soviet intellectual rally. He won a standing ovation upon entering the room, and gave a speech on the responsibility of the individual writer, which those present remembered as being characterized by long, impressive silences. He appeared to many to be ill, and his meeting with Tsvetaeva, so longed-for nine years earlier, now became an almost formal event, in which he was asked to provide information about the Soviet Union. Tsvetaeva's husband and children, worn down by relentless poverty and eager to believe the newly efficient Soviet propaganda, were pressing her to go back; she herself had recently written a poem called "Homesickness." Trapped in his official role and in a state of shock and supreme fatigue, Pasternak told them very little. Of what he did say, it is impossible to imagine the tone of voice—some historians believe that Tsvetaeva missed the irony—with which he confided, "You will come to love the collective farms."

It was not Pasternak who was responsible for Tsvetaeva's return to the Soviet Union, however, but Sergei Efron. He had found work, at last, with a political organization that was, as it turned out, a tool of the NKVD, the Soviet secret service. Drawn into their activities, Efron in 1938 was implicated in several murders (including that of Trotsky's son) and was arrested by Paris police. Released pending further questioning, he escaped, through Spain, to the Soviet Union. Ariadna Efron had already gone, and Tsvetaeva, left with thirteen-year-old Mur, found herself not merely isolated—she was used to isolation—but the victim of furious suspicion and disdain. Nina Berberova tells of Tsvetaeva standing alone at the funeral of a friend, a Russian poet; no one would go up to her and speak. No one could believe that she had not known of Efron's treason. In June 1939, after securing safekeeping for her manuscripts and her precious letters from Rilke, the woman who had once written so confidently, "I would have been able to say to Orpheus: 'Don't look back!' " and had faulted the blindness of Eurydice's love packed her bags and followed her husband into Hell.

Tsvetaeva's return to Moscow, with Mur, went unremarked. She came not as a renowned poet but as a former supporter of the White Guard and the wife of a disgraced Soviet agent. She and Mur were sent to an outlying house reserved for such pariahs and their families, where they were reunited with Sergei and Ariadna. Tsvetaeva later wrote in her diary, "The overtone and the undertone of everything: horror." In August, Ariadna was arrested; she went on to spend seventeen years in northern prison camps and internal exile. Sergei was taken away that fall. He was shot, as far as can be determined, in 1941.

Evicted, despised, without work or money, and with Mur as well as herself to support, Tsvetaeva appealed to Pasternak, who refused at first to see her—"It's dangerous," he warned a friend—but arranged for her to receive translation work and meals at a Writers' Union club outside Moscow. When the price of the meals was raised beyond her means the following spring, she began searching for a room in the city that had once been her own. "For a year now," she wrote to Pasternak, "I've been looking for a hook. . . . For a year I've been trying on death. Every way is ugly and frightening. . . . I don't want to die, I want not to exist."

Yet there were bouts of what amounted to a kind of normalcy even in this period. She found a place to live (thanks to Pasternak), she published a poem (albeit an old one), and she was even asked to prepare a manuscript of her selected works for an official Soviet publishing house. True to form, however, she only brought on official wrath, by including in her manuscript, as Viktoria Schweitzer has shown, none of the poems—such as "Homesickness," a wistful admission of an émigré's yearning for the Russian landscape—that might have allowed her to seem a grateful prodigal returned. Instead, she began with a poem written for Sergei in his White Guard days, which she dedicated, openly and dangerously, "To S.E."

And after all the years of distant adulation she was introduced, at last, to Akhmatova. The two women sat talking together for hours, although they were united now more as terrified mothers, it seems—Akhmatova was in Moscow to petition for her son's release from prison—than as poets. (When each recited from her recent work, the result was mutual incomprehension.) Akhmatova could not bring herself to present Tsvetaeva with the one poem she had finally written for her, in March 1940, entitled "Belated Reply," in which Tsvetaeva's famous Moscow bells have become funeral bells, and in which, taking on Tsvetaeva's voice, Akhmatova moans of how "An abyss has swallowed my loved ones / And my parents' home has been pillaged."

In June 1941 Germany attacked Russia, and a month later Moscow was being bombed. Pasternak politely deflected Tsvetaeva's request to let her and Mur stay at his dacha outside the city, but he saw her off on a boat taking evacuees to Tatarstan. She and Mur arrived in the tiny town of Yelabuga on August 17 and were given a room requisitioned from a local family, an old couple who lived with their grandson. Looking for work, Tsvetaeva ventured to the larger town of Chistopol, where a settlement for writers better regarded than she had been established. After much petitioning and outright begging, she secured a residence permit and the possibility of a job as a dishwasher in the writers' canteen.

While in Chistopol, in despair, Tsvetaeva was recognized and taken in by sympathetic admirers. Assured of the little circle's approval and asked to recite from her work—they requested the old "Poems to Blok"—Tsvetaeva chose instead to recite "Homesickness" but was too shaken to get through to the end. Slipping away, she returned to Yelabuga. Three days later, while Mur and the others were out, she found her hook; actually, it was a nail, "Used for the yokes of horses, too low to / reach for, still less hang from," if a poem written by Yevgeny Yevtushenko decades later is to be trusted. (Yevtushenko concludes, "Remember: there is only / murder in this world. Suicide has no existence here.")

It was the old woman of the house who came home first and found her. It was Sunday afternoon, August 31, two years after Ariadna Efron's arrest. A note to Mur, found in Tsvetaeva's apron pocket, asked his forgiveness, and added, "If you should ever see Seryozha and Alya, tell them that I loved them until the last minute." He never saw either one; drafted into the Soviet army, Mur died in battle in 1944. As for "Seryozha and Alya," documents newly discovered in the archives of the Soviet secret police, published by Irma Kudrova, indicate that both father and daughter had been deeply involved in informing activities, and had worked to implicate each other and Tsvetaeva herself. The material suggests that Tsvetaeva's realization of this ultimate corruption may have led her to her final act.

The old woman of the house was asked about her strange boarder many times in later years, first by lone pilgrims and later by literary societies and school groups and biographers, until the house became something of a museum and she was tired of trudging to the gate. She replied to the most frequently asked question by saying, rather quizzically, that there had been plenty of food left at the time—a sack of "rice and pearl barley and different kinds of meal"—and that surely "she could have kept going longer," that "it would have been time enough when they'd eaten it all."

Nadezhda Mandelstam, whose memoirs of life and death under Stalin exhibit an exacting discrimination among degrees of tragedy, claimed to know of "no fate more terrible than Marina Tsvetaeva's." Russia's foremost widow was weighing the fates of those poets who, with their burden of witness and conscience, have stood out, from so many victims, as the Soviet Union's exemplary dead. ("Why do you complain?" Osip Mandelstam once chided his wife. "Poetry is respected only in this country—people are killed for it.") By the logic of this system, it may be seen as the last indignity that Tsvetaeva was left to manage her death entirely on her own. The lack of any openly directed state persecution—such as that which killed Mandelstam and silenced Akhmatova and finally threatened (and canonized) Pasternak—denied her even the compensation of public martyrdom, that final proof of dangerous importance for which Western writers once guiltily envied Soviet suffering. The revival of Tsvetaeva's literary fortunes offers little by way of political instruction, except in her aspirations beyond all ideology and public poses; her victory is in the intense, womanly privacy of the poems themselves. Yet even deprived of the full force of her poetry, awaiting the translator who will reimagine her powers for English readers, Tsvetaeva draws us deep into her presiding myths, her life as large and unnerving and painfully exalted as the lives of the Phaedras and the Ariadnes for whom she tried to speak.

Twilight of the Goddess

Ayn Rand

"This relationship is sexual or it's nothing," Ayn Rand warned her leading disciple in 1955, when their affair and his qualms were new. In his recollection of the scene, her words betrayed neither jealousy nor fear nor any of the other weaknesses suggested by a woman's setting such an ultimatum, although Rand was fifty years old and he, her most ardent reader, only twenty-five. But then, few American novelists since Harriet Beecher Stowe have had the force or the desire to turn readers into disciples, to shape political or religious or moral convictions. By the respectable terms of the modern literary profession, novelists do not preach. And, in fact, there has probably not been a less respectable novelist among the irrefutably enduring writers of our time than Ayn Rand: philosopher queen of the best-seller lists in the forties and fifties, cult phenomenon and nationally declared threat to public morality in the sixties, guru to the Libertarians and to White House economic policy in the seventies, and a continuing exemplar of Wilde's tragic observation that more than half of modern culture depends on what one shouldn't read.

Almost two decades after her death, Rand's books still sell more than three hundred thousand copies a year. *The Fountainhead,* her slow-building blockbuster of 1943, was made an honorary Book-of-the-Month Club selection fifty years after its publication, in tribute to the public's continuous, unabating demand. But the ever-renewing size of Rand's audience is only a part of the story. At a *Fountainhead* anniversary banquet held by the Ayn Rand Institute, nearly two hundred people paid a hundred and twenty-five dollars each to listen to excerpts from Rand's private letters, and to watch each other bid more than five thousand dollars for her blue-green metal ashtray and matching lighter (the last-minute addition of two cigarettes marked with a dollar sign, a famous Rand prop, doubtless drove up the price) and twenty-five thousand dollars for the manuscript of her last speech, made in 1981 to the National Committee for Monetary Reform, for which Rand herself had been paid in gold. Throughout the festivities, responsibly conservative business executives, teachers, secretaries, lawyers, and a scattering of college students who'd been barely old enough to read at the time of Rand's death discussed the principles of heroic individualism by which she had taught them all to live.

"It is true that fiction is a much more powerful weapon to sell ideas than non-fiction," Rand noted in 1944, in one of the many pronouncements on her art now available to all in the engagingly hale and generous, if none too private, *Letters of Ayn Rand.* Edited by Michael S. Berliner, the executive director of the Ayn Rand Institute, the selection generally keeps to the official portrait of an indomitably spirited ideologue: in this 668-page record, there is no trace of intellectual doubt or growth, or, for that matter, of the infamous love affair that tore apart the Objectivist movement, which had for years officially represented her philosophy. Yet even within the largely impersonal confines of the *Letters,* a boundlessly brave and yearning and often foolish woman manages to peer through, with the repeated claim that she is altogether a triumphantly rational being—nothing more or less than the sum of her ideas.

It is with pride that Rand identified herself, in a letter of 1944, as "the chief living writer of propaganda fiction." Although she was to become less satisfied with that appraisal, Rand's entire body of work—four novels, assorted plays and screenplays, hundreds of essays and articles—remains a declaration of principles about men and women and money and the life worth living. And her life, like those of her disciples, was meant to be the indisputable proof that her goals could be achieved. The threat she hurled at her lover that day in 1955 drew its assurance from a point of her essential doctrine: "If we are not a man and woman to each other, in the full sense—if we are merely disembodied minds—our philosophy is meaningless."

At the time, Rand was mired in a twelve-year struggle to complete her last novel, *Atlas Shrugged,* a manifesto of her "strictly American" philosophy of "rational selfishness," which would weigh in at well over a thousand pages. Rand entertained the vast ambitions of a late-Victorian novelist, or a Russian one: when she was just beginning *Atlas Shrugged,* she promised a reporter that her new book would combine "metaphysics, morality, economics, politics and sex." She meant to prove that absolute rules of behavior could be derived from objective reality; that there was no conflict between ethics, practical ambition, and having a fabulous time. All this omniscience and optimism she hammered into a sprawling triple-decker romance that was by turns melodramatic and speechifying, titillating and edifying—a best-seller in a tradition so nearly extinct that it seemed new. Not since the popular novels of almost a century before, bent on refutations of Darwin or God and offering what George Eliot called "a complete theory of life and manual of divinity, in a love story," had there appeared so vividly accessible and reassuring a guide for the cosmically perplexed. As late as 1991, the Library of Congress found that a majority of Americans surveyed named *Atlas Shrugged* as the book that had most influenced their lives, after the Bible.

Late in her life, Rand maintained that the impetus behind her long toil was allegiance not to a reforming politics or a moral abstraction but to a strictly private and sensual vision. "The motive and purpose of my writing is *the projection of an ideal man*" is how she began a 1963 essay called "The Goal of My Writing." All the rest had followed from the need to provide a world in which this ideal could draw breath. "My purpose, first cause and prime mover is the portrayal of Howard Roark or John Galt or Hank Rearden or Francisco d'Anconia *as an end in himself,*" she went on, listing the major heroes of her fiction—all giants of *Homo erectus* for whom only a vacated Heaven could clear sufficient height and an unrestrained capitalism offer sufficient expanse for their gifts of conquering and possession. There is a profound unity of intent behind Rand's stupendous fantasies about an absent God, the false ideals of Communism, and the beauty of naked men.

■ ■ ■

The most driven of American literary anti-Communists was born in St. Petersburg in 1905. Her name then was Alisa Rosenbaum. She was the oldest of three sisters, physically small and dark, with enormous eyes and a tendency to stockiness; a solitary child, she was more intelligent and far more serious than anyone she knew. Her father, Sinovi, was a Western-leaning political thinker—a pharmacist, whose profession had been determined by the quotas set for Jews at the city's university. Alisa was present in her father's shop the day Bolshevik soldiers broke in and affixed a red seal to the door, declaring his life's work the property of the Soviet people. Decades later, she recalled his look of "helpless, murderous frustration and indignation." She was already pledged to being a writer, and had been pouring out stories since the age of nine. Now she had a subject.

She began to make lists of themes for future use, and to keep a diary of her ideas. "Today, I decided to be an atheist," reads an entry she made at thirteen—in a consciously feminine transfer of loyalty from an uppercase to a lowercase god. "I had decided that the concept of God is degrading to men," she explained years afterward; men should have nothing placed above them. Already, she was working out a kind of remedial atheism that would do away with the insupportable pain of cosmic loneliness—at least, for women and imperfect men—by allowing the lesser mortals to retain someone to worship.

This eagerly avowed hero worship—she would one day contrast it proudly with Communism's "zero worship"—was based not on any man she had ever seen but on the books she had read. In her passion for Hugo's Jean Valjean and, especially, Nietzsche's Zarathustra, Alisa Rosenbaum was a true child of the last imperial decades, when a cult of the *Übermensch* overtook Russian thinking and spread in an epidemic of bastardized and novelized Nietzsche, sporting a contingent of heroes generically arrogant and oversexed and vaingloriously free. Ayn Rand's lifelong theme of the well-muscled individual versus the lumpen masses was, at the start, merely a continuation of a cultural opposition set out decades before, with the failed liberal ambitions of the Russian mid-century. It was with the Revolution that this romantic persuasion became a political necessity.

After the confiscation of the Rosenbaum shop, the family spent three years in the Ukraine, waiting for the Revolution to fail. In Odessa, they fought off bouts of starvation and scurvy. Meanwhile, Alisa attended high school and studied American history ("I thought: "*This* is the kind of government I approve of"). In 1921 Sinovi decided to take his family home, despite his wife's pleas to follow the White exodus from Russia. He was convinced that reason would prevail or be imposed from the West, that the Bolshevik seizure could not last. Among his daughter's enduring memories of Odessa was an official "Week of Poverty" during which Red soldiers came to remove any excess possessions and went away with the family's soap; but there was also the unexpected pleasure of her first job—she had thought she would hate it—teaching Red soldiers to read. Before leaving for Petrograd, Alisa burned all her diaries and theme books lest they fall into enemy hands.

She was sixteen years old when the family returned, and she fell in love for the first time with a real person, precisely because "he didn't look real," as she put it: "He was so perfectly good-looking." An engineering student named Leo, he was tall and angular and aristocratic; she saw him at a party and couldn't stop staring. The next time they met, he walked her home, and by the time they arrived she was "madly and desperately in love." He dated her for only a short time, and when he stopped "that was the most prolonged period of pain in my life," she recalled several decades later. And there was no consolation. She was not "social" (to her mother's despair) and disliked those who were (her mother above all). She was like Leo, in fact—incapable of the ordinary, exquisitely alone, marked for a special destiny—except that none of this showed in the way she looked.

That fall, she enrolled as a history major at the University of Petrograd. She took the required courses in dialectical materialism but received her meaningful education at the movies. Foreign movies, Hollywood movies: if Marxism was the new state religion, movies were the new opium of the people, and the brightest stars on the Soviet horizon were Chaplin, Pickford, and Fairbanks. (Eisenstein's *Potemkin* was bounced from its theatre run by Fairbanks's rather ideologically similar *Robin Hood.*) By 1924, her senior year, Alisa Rosenbaum was going to the movies every night. For a single long shot of the New York skyline she would sit through a feature twice.

In the fall of the next year, a letter arrived for the Rosenbaums from distant relatives, the Portnoy family of Chicago. The Rosenbaums had helped to pay the Portnoys' steerage when they fled Russia in 1889, and the Portnoys' children were now in a position to return the kindness. Alisa was the only member of her family to apply for a visa, over her father's objections; Sinovi was still awaiting the imminent collapse of the Soviet state. She began to study English, and she enrolled in a film school that had opened in Petrograd, in order to learn to write for the movies in the new world that movies had taught her to see. When the visa came through, her parents gave her a farewell dinner. On that hope-filled occasion early in 1926, one of the guests made a request that she held on to for years: "If they ask you, in America—tell them that Russia is a huge cemetery, and that we are all dying slowly." She promised she would tell them in America.

She missed the skyline because of the fog, and didn't realize she had arrived. Awaiting admission in February 1926, she rejected plain English Alice for exotic Ayn—after a Finnish writer she'd never read—and stepped onto the Hudson pier, on a snowy evening, the not quite transformed Ayn Rosenbaum. In her arms she carried the battered mechanism of her full metamorphosis—the Remington Rand typewriter from which she would, within weeks, take the rest of her name and begin to recompose her life.

Standing at last before the hard, soaring facts of the city, she resolved to write a novel with the skyscraper as its theme. In *The Fountainhead,* which appeared seventeen years later, a wealthy newspaper magnate expounds on just such a sight as greeted the author that first dim evening. "I would give the greatest sunset in the world for one sight of New York's skyline," Gail Wynand declares from the deck of his magnificent yacht. "Particularly when one can't see the details. Just the shapes. The shapes and the thought that made them. The sky over New York and the will of man made visible. What other religion do we need?"

In fact, there was nothing in Rand's vision of the modern Civitas Dei that hadn't been intended by the architects of modern capitalist theology, who were well aware that their works were the first to rise above the city's church towers. Still dominating the skyline in the year she arrived was the Woolworth Building, a sixty-story Gothic tower that had been presented to the world as a "cathedral of commerce" and described as "a battlement of the paradise of God" in a brochure issued shortly after its completion. For Rand, moved to a rare anatomical acknowledgment, it "stood out ablaze like the finger of God."

She spent just a week in Heaven before she had to join her relatives in the purgatory of Chicago. The Portnoys were to remember her as a thoughtless young woman who refused to speak of Russia or her family and who kept them up all night with her typing. She seems to have hardly remembered them at all. She was off again by midsummer, having completed her first story in hesitant English, about unrequited love. She had also completed four screenplays—an easier form for a foreigner to handle in the days before talkies. Most notable was one about a "skyscraper hero" who leaps from building to building by means of a parachute. She was taking it to Hollywood.

And Hollywood was ready for her with a real-life scenario of about equal credibility. On the day after she arrived, she failed to get a job in the screenwriting department of the DeMille studios but met Cecil B. himself in an open roadster by the gate. He saw her staring. She told him she had just arrived from Russia. He told her to get in, and drove her to the location of *King of Kings,* up in the epic hills of a fresh-painted Jerusalem. A few days later, she had a job as an extra and was making more money than ever before in her life. She took a room in the famed, girls-only Studio Club, where she must have cut an unlikely figure among the aspiring starlets, with her dark and heavy foreign clothes, her obliterating accent, her square-jawed solemnity. Although the rent of ten dollars a week included two full meals a day, she was always starving; "It took several years," she recalled, "for the constant hunger to disappear."

It was on the long streetcar ride to the set, one groggy dawn, that she looked up to see a stranger who was tall and angular and unmistakably Zarathustran: "It was love at first sight. . . . I have never seen a face that would fit my view of the ideal man quite as well." He got off at her stop; he had a bit part in the same film. The next time she caught a glimpse of him, he was wearing a toga and sandals laced to the knees, and he looked even better. Three days after that, she tripped him on the road to Calvary and they were introduced.

But the day after that he was gone, his part finished. Although she knew his name—Frank O'Connor—she couldn't trace him anywhere. For months, she was stricken; at the Studio Club, she became known as the Russian girl who cried in her room. Only work went well. DeMille had given her a chance as a junior writer, and in the summer of 1927, while researching a picture in a library, she ran into Frank again. Their friends would always think them a mismatched pair. Despite his looks, O'Connor was an extremely gentle man, kind and unambitious and not overly interested in anybody's burning ideas. When they were married, in April 1929, it was generally known in their circle that he was going through with it as a favor, because her visa was about to expire—and that she was madly in love.

A citizen at last, she supported her husband's occasional acting career with a job in the wardrobe department at RKO. Nights, she worked on becoming a real American writer, refining her grammar and studying the plot ingenuities of O. Henry. Her first sale, to Universal Pictures, was a screenplay about a beautiful woman—the part was intended for Marlene Dietrich—who comes to a Soviet prison island as a whore for the commandant, known as the Beast. ("You may find I deserve the name," he warns her. She replies, "You may find I like it.") In reality, she is the noble wife of a handsome young political prisoner and has come to help her husband escape. But the Beast falls in love with Beauty and realizes that he, too, must escape, not only the island but the larger prison of the Soviet Union. Love—the intractable human need for sole possession of the beloved—has taught him the folly of Communism. The screenplay was not produced, but the money Rand was paid for it enabled her to quit the wardrobe department and devote herself to her serious writing: a novel, set in Petrograd, about a beautiful woman caught between her arrogant anti-Communist lover and a handsome Bolshevik whose desire to possess her shatters his false ideals.

▪ ▪ ▪

> "I know what you're going to say. You're going to say, as so many of our enemies do, that you admire our ideals, but loathe our methods."
>
> "I loathe your ideals."

We the Living, Rand's first novel, was completed in 1933. A triumph of English as a second language, if not of English itself, it is a proficiently staged melodrama enlivened by fervent ideological dressing and undressing. The heroine, an ardent anti-Bolshevik ("I know no worse injustice than the giving of the undeserved"), has been mistress to both men not through wantonness but through nobility, to save the one she truly loves. Despite such clear enticements to publication, the novel received a steady series of rejections, which the author resolutely blamed on her work's being "too intellectual" and "concerned with *ideas.*"

In late 1934 Rand and her husband set out for New York to monitor the Broadway production of *The Night of January 16th,* her first commercial success. It was a courtroom drama with a kick: the audience got to vote on the verdict. The play earned her the money to settle on Park Avenue and begin a new novel, and the reputation to attract a publisher for the one she had already finished. In the fall of 1935, the Macmillan Company paid two hundred and fifty dollars for the rights to *We the Living,* overriding the protests of a manuscript reader, Granville Hicks, who was best known as a spokesman for New York's spectacularly literate Communist Party.

Entering the American thirties must have been for Rand like falling into a looking-glass world, in which the best and most lauded minds insisted that the Soviet cemetery she had vowed to expose was a working model of paradise, and in which all testimony to the contrary was disbelieved or despised. Edmund Wilson's 1931 condemnation of "the monstrosities of capitalism" remained a touchstone for the intellectual decade. In 1936 reviews of *We the Living* were juxtaposed, by chance, with adulatory accounts of Granville Hicks's new biography of John Reed, the Harvard boy who had travelled in the opposite direction from Rand and been buried with honors in the Kremlin wall: "a revolutionary out of the tenderness of his heart, not out of hatred," the *Times* reviewer wrote. By contrast, the paper found Rand's novel "slavishly warped to the dictates of propaganda," and her considerable narrative skills undermined by "the blind fervor with which she has dedicated herself to the annihilation of the Soviet Union." *The Nation* decried her attempt to "puncture a bubble with a bludgeon," in a notice followed by ads for tours of the Soviet Union ("high standards of comfort" guaranteed). If Ayn Rand had set out to become the century's Harriet Beecher Stowe, her task was all the more daunting. She was not morally lovable, and she was on the wrong side.

We the Living quickly sank from sight, and in her fiction Rand never again wrote directly about the Soviet Union. This was not just a response to failure; the more immediate terrors of New Deal legislation had brought her to a broader and even more desperate view of the collectivist threat. In 1937 she paused in the labor of plotting her big new novel to write a brief science fiction parable entitled *Anthem,* which was clearly (if unadmittedly) based on Evgeny Zamyatin's *We* (1924), the first Soviet novel to be banned in its homeland and subsequently the inspiration for Orwell's *Nineteen Eighty-four.* Like its prototype, Rand's *Anthem*—originally titled "Ego"—is set in a future Communist state of numbed and numbered citizens, but it features a pair of purely Randian lovers struggling to rediscover the lost first person singular: "We love you" does not quite express what they are longing to say. Although *Anthem* appeared in England in 1937, it was rejected by every publisher the author approached in her own country.

"To understand how I felt, you have to know what kind of books *were* being published in the thirties, and hailed as serious literature," Rand later remarked. She meant, of course, books by celebrated writers like Steinbeck and Farrell and Dos Passos, books rooted in social collapse and disillusion, without heroes—clumsy books that were on the right side. The only contemporary novel that she passionately admired was a drama of social revolution which told essentially the same tale as *We the Living,* as she saw it: "the romantic conflict of a woman who loves a man representing the older order, and is loved by another man, representing the new." So runs Rand's summary of *Gone with the Wind,* a first novel released with great fanfare by Macmillan in the same season that *We the Living* was dropped into the void. Rand watched closely as the public trajectories of the books parted. She compared Margaret Mitchell's book with her old favorite, Hugo's *Les Misérables,* which she reread and minutely analyzed, taking apart the mechanisms of plot and theme and structure in the hope of assembling just such a great, whirring narrative machine of her own—this time, one that would fly.

▪ ▪ ▪

> "Why did you become an architect?"
> "I didn't know it then. But it's because I've never believed in God."

Howard Roark decided on his vocation at the age of ten, when he first noticed that the earth could do with some remaking. This is the only background information we are given about the hero who springs full-blown onto the first page of *The Fountainhead.* No family, no mistakes, no uncertainties; this is the ideal man, hewn from the Nietzschean rock face of the author's will to dream. When the book opens, Roark is standing naked on a rather more Freudian rock face, surveying the surrounding granite cliffs as they stand "waiting for the drill, the dynamite and my voice; waiting to be split, ripped, pounded, reborn; waiting for the shape my hands will give them." He has just been expelled from architecture school for being entirely original and entirely unwilling to accept the judgment of others. (" 'Why do you want me to think that *this* is great architecture?' Roark pointed to the picture. . . . 'It's the *Parthenon!*' said the Dean.")

For 754 pages, his glance ever steady above the abyss of the unconscious, the Ideal Man stands against the forces of mediocrity and collectivism, represented—this is, after all, America—by newspaper critics and modern artists and a large cast of envious parasites. Commissions are few, but he builds a temple in which men feel too proud to kneel; a house so natural it seems to grow out of the cliff on which it stands; and an apartment building that sets spectators musing on the phrase "in His image and likeness." In the book's dramatic climax, when Roark's design for a housing project for the poor is altered in execution, he simply blows it up. On trial for the crime, he defends himself with an eight-page lecture on world history and its contemporary perils, declaring, "I recognize no obligations toward men except one: to respect their freedom and to take no part in a slave society." He is acquitted instantaneously. On the last page, we see him standing high atop his latest creation—the tallest skyscraper in New York—with his clothes on but with the world firmly restored to its initial location at his feet.

It is surely gratuitous to point out that the author suffered from an edifice complex. Or that in a Freudian age the stark city tower, stripped of its Gothic encrustation, was a most accommodating symbol of Eros emerging from the carapace of Belief. Roark's apotheosis leaves him just where he belongs: an Art Deco god affixed to a skyscraper, an overscaled figure carved into language as heavily outlined and flattened and abrupt as a thirties architectural relief. The simplified, monumentalizing style that Rand called "romantic realism"—presenting "life, not as it is, but as it could be and should be"—had its only counterpart, ironically, in the officially sanctioned Soviet bombast of Socialist Realism: in its high legibility, its hectoring force, and its insistence that there were still giants among men, battling for one side or the other.

In 1937 Rand had written a worshipful letter to Frank Lloyd Wright and asked for an interview in regard to her projected novel. Although Wright turned her down, his buildings and goals, and even his much publicized personality, underlie her image of the architect as isolated genius and Lawrentian hero. ("He lived from first to last like a god; one who acts but is not acted upon." That is not Rand but Lewis Mumford on Wright.) The scrupulous novelist put in six months typing and filing at an architectural firm, learning the lesser properties of granite. But her book is not about architecture, as she had to protest—however astonishingly—many times. *The Fountainhead* once again tells Rand's only story: the individual versus the masses, played out at her private crossroads of political and erotic retribution.

The manuscript was rejected by twelve publishers before an editor at Bobbs-Merrill staked his job on the still unfinished book. *The Fountainhead* appeared in May 1943, substantially cut (to its present form) because of the wartime paper shortage, but with its author's purposes intact: she had refused to eliminate a word of Roark's climactic speech. Reviews were spectacularly divided, ranging from a Sunday *Times* tribute to "the only novel of ideas written by an American woman," who was also "a writer of great power," to Diana Trilling's warning in *The Nation* that "anyone who is taken in by it deserves a stern lecture on paper-rationing."

The first printing, of seventy-five hundred, took several months to sell. A second printing, of twenty-five hundred, followed in the fall, and also moved slowly. But the sales department at Bobbs-Merrill began to hear tales like the ones Rand exults over in replies to her fans: a manufacturer of fishing tackle in Kalamazoo who had put up his own money to advertise the book; a public librarian who had personally built her collection of copies from two to thirty. *The Fountainhead* was fast becoming "the greatest word-of-mouth book" that any publisher had seen. And it was being read not just widely but intensely: underlined, pressed on friends, argued over, reread. Impassioned readers were emerging from the largely abandoned American class of thinking nonintellectuals—readers who enjoyed the story and were excited or flattered just to be put in touch with provocative ideas. If you didn't find the book ridiculous from a literary point of view, chances are you found it immensely stimulating. Many readers were inspired by the author's claims for an absolute individualism—that eternal self-dramatizing romance of the American male, from Thoreau's cabin to the last boozy choruses of "My Way." As for the romance of the American female, *The Fountainhead* contained the most stimulating example yet of that essential component of the best-selling "woman's novel": the rape.

Rand first conceived of her exquisitely perverse heroine while travelling through Virginia on her way from Hollywood to New York and passing, in quick succession, a chain gang and an old white-columned mansion. The result was a "chatelaine of the countryside" called Dominique Francon, delicate and pale as glass, a virgin who believes herself frigid despite a longing "to be crushed"—by trees, generally—and a compulsive tendency to visit a nearby granite quarry where "she knew she would face a gang of workers." Rand began writing her book with the scene in the quarry where Dominique spies "the abstraction of strength made visible," the man down in the pit:

> He stood looking up at her; it was not a glance, but an act of ownership. She thought she must let her face give him the answer he deserved. But she was looking, instead, at the stone dust on his burned arms, the wet shirt clinging to his ribs, the lines of his long legs. She was thinking of those statues of men she had always sought; she was wondering what he would look like naked. She saw him looking at her as if he knew that. She thought she had found an aim in life—a sudden, sweeping hatred for that man.

For those feminists who have decried the big scene that is on the way—the scene that Susan Brownmiller set as the central shame of modern fiction in her 1975 study *Against Our Will: Men, Women and Rape*—it is worth emphasizing just who does the initial sizing up and mental undressing and desiring. Rand defended her sexual dramatics with a clear distinction: "Literal rape would be contemptible and disgusting and unthinkable to any hero of mine." As for what happens in her novel, "if that was rape—it was rape by engraved invitation," a notion that only makes the issue murkier or, for many women, more infuriating.

The scene that brought Brownmiller to indict Rand as "a traitor to her own sex" begins when the insolent, work-stained quarryman—in reality, of course, Howard Roark, certified genius and the heroine's future husband—enters her white-and-crystal bedroom through its open French doors. After ten pages of narrative foreplay, the actual event is brief. He lifts her "without effort," she beats on his chest but makes no sound, he throws her on the bed. "She felt the hatred and his hands; his hands moving over her body, the hands that broke granite. She fought in a last convulsion. Then the sudden pain shot up, through her body, to her throat, and she screamed. Then she lay still." The scraping together of these fleshless bones put Rand into a rare literary swirl, her distinctively short hard sentences giving way to the "thing she had thought about, had expected, had never known to be like this, could not have known" school of flushed-pink prose. The result incited a countryful of readers to linger and to dog-ear so much that Brownmiller, taking out a library copy, found that it fell open to just the scene she was looking for.

The Fountainhead hit the best-seller list in 1945 and stayed there for more than half the year. To a friend who doubted whether the severely independent-minded author could care for such meaningless acclaim, Rand replied that on seeing the first list "I started screaming—literally and loud." This was probably the most buoyant time in her life. The public enthusiasm almost compensated for the wounding critical snubs; in fact, in her somewhat awkward success at mass-marketing individualism, the critics could be used to reassure her that her readers—however legion—were still "the exceptions, the prime movers, the men who do their own thinking." And then she was going back to Hollywood in triumph, to be courted by movie stars vying to play the hottest roles since Scarlett and Rhett.

At the urging of Barbara Stanwyck, Warner Brothers had bought *The Fountainhead* early, in 1943, closing off the chance that Clark Gable, at MGM, might play Roark; Gable later told Rand that he'd "raised hell" with his studio for not protecting his interests. As it turned out, Stanwyck didn't get the role of Dominique, and Gable had a lot to thank his studio for when the King Vidor movie finally appeared, in 1949, an unintended comedy played at the wrong speed and in a collision of styles, from arch expressionism to dismaying earnestness: Gary Cooper ambling through the cabinet of Dr. Caligari. Rand had fully recognized the difficulties that any attempt to "humanize" her characters would introduce. "Heroes do not say 'Gee whiz'—nor any equivalent of it," she warned the producer in an unavailing plea for the all-out stylization of the silent films that had inspired her in the first place. Her book could not be normally cast and filmed, she knew, because there were no people in it.

In Hollywood, inevitably, she was drawn into industry politics. She went to Washington in 1947 to testify before the House Un-American Activities Committee about anti-Americanism in movies, but the ineptitude of her allies proved as disturbing as the derision of her opponents, and she returned home withdrawn and bitter. Nor did she find much there to cheer her. She was living in isolation on a ranch in the San Fernando Valley, at work on a new novel of staggering ambition. Because the famed worshipper of self-sufficiency and technology had never learned to drive, she was dependent for every move on her husband, who had discovered his true vocation in the tending of their flower gardens, and his pleasure in keeping to himself. She was already using amphetamines—a low but daily dosage—to keep her weight down and her energy up, and she now began to be dependent, as well, on a series of young *Fountainhead* acolytes who arrived at the ranch in search of further wisdom.

In March 1950 a nineteen-year-old UCLA psychology major named Nathan Blumenthal earned an invitation to visit after sending Rand a letter full of philosophical questions about her work. He was surprised to find his favorite author neither as thin nor as elegant as her book jacket photograph suggested; she reminded him of his own Russian-Jewish cousins. Blumenthal himself was tall and—to judge from photographs—somewhat pudgy, and his chief act of heroism consisted of his having read *The Fountainhead,* as he told her, "almost continuously" since he had borrowed his sister's copy at fourteen, after hearing her reading the sexy parts out loud with her friends. But the parts that mattered to *him* were the hero's firm rejections of society's judgments and rewards. He and Rand sat up talking all night, while Frank served them coffee and cookies. Rand soon found her visitor to be a "budding genius" and, eventually—he quoted her as saying—"the realization of everything I write about."

Six hundred and fifty pages into *Atlas Shrugged,* the intrepid heroine Dagny Taggart has crash-landed the plane she's been flying in pursuit of a mysterious opponent, who is bending over her tenderly as she regains consciousness. She looks up:

> The shape of his mouth was pride. . . . The angular planes of his cheeks made her think of arrogance, of tension, of scorn—yet the face had none of these qualities, it had their final sum: a look of serene determination and of certainty. . . . It seemed to her for a moment that she was in the presence of a being who was pure consciousness—yet she had never been so aware of a man's body.

Dagny has twisted an ankle in the crash, and her obliging opponent lifts her (effortlessly) in his arms and totes her off to see the secret world he has brought into being: a Utopia populated entirely by heroes, by men (and three women) who have been the country's leaders in science and industry and business and art. They have gone into hiding—on strike—in order to make a parasitic society aware that its existence depends on their talents, and to force the repeal of laws and regulations that constrain them. Self-sufficiency is the supreme good in Utopia, and everyone is strong enough or smart enough or rich enough to thrive. The symbol of the cross has been replaced with the dollar sign; gold is the official currency, and the verb "to give" is forbidden. "They had known that theirs was the power," proclaims John Galt, the group's leader and the man Dagny chased until he caught her. "I taught them that theirs was the glory."

Same story, new symbol: the individual as fountainhead has become the individual as motor of society. John Galt was inspired to begin "The Strike"—Rand's original title—when the motor factory in which he worked as an inventor went over to a profit system described as "from each according to his ability, to each according to his need." Rand doesn't actually mention Marx—or, for that matter, the Soviet Union—but as political science fiction her account of the fate of the Twentieth Century Motor Company is stunningly accurate. As emotional science fiction, the novel contains almost nothing related to human experience, except on those occasions when a woman's voice breaks through on the subjects of her pride in her work and her sexual longing.

Rand spoke laughingly of Dagny Taggart as "myself, with any possible flaws eliminated." Like all Rand heroines, Dagny is an Aryan goddess, lithe and fair and "arrogantly, ostentatiously patrician." (Rand said that she had Katharine Hepburn more or less in mind.) She is also a model of worldly competence, running a railroad single-handed. Of the other women we glimpse in Galt's Utopia, one is an actress whose chief commodity is her ravishing beauty, and the other a writer—she makes just a cameo appearance—with dark and dishevelled hair and enormous eyes, who leaps to her feet when Galt approaches and stares after him with a look of "hopelessness, serenely accepted." She inspires in Dagny the only stab of jealousy the perfect girl has ever known.

For Rand, social isolation and disappointment gave way in the early fifties to the constant company of Nathan Blumenthal and his girlfriend, Barbara—another *Fountainhead* votary, whom he had brought along on his second visit to meet the revered author. Barbara was quiet and serious, and was physically "in the Dominique style," Rand noted; though she was clearly not Nathan's equal, Rand approved of the match. When "the kids" moved to New York, in the summer of 1951, Rand and her husband followed. Frank missed the ranch and his flowers terribly, but he eventually found a part-time job with a florist and took up painting in his ample spare time.

In 1953 the O'Connors served as matron of honor and best man at the

Blumenthals' wedding. The couples were neighbors, living in the East Thirties, and every Saturday night a group of young disciples would gather in Rand's living room to hear new parts of her work in progress and to discuss her ideas. The Class of '43—as Rand called the group, after *The Fountainhead*'s year of publication—included, most illustriously, the young economist Alan Greenspan, whose appointment as President Ford's chairman of the Council of Economic Advisers some two decades later was announced in *Newsweek* under the heading FUNDAMENTAL FOUNTAINHEAD. (Rand attended the White House swearing-in.) Nearly all the disciples were in their twenties, and they were were exhilarated to feel themselves a part of history. Some expected a government conversion to full free-market capitalism as an immediate result of the publication of "The Strike."

But Rand was writing a sexual manifesto as much as a political one, based on a somewhat irregular application of the theory of the unity of body and spirit. A man's physical desirability reflected his superior mind and character, in Rand's view, and a woman's superior mind and character made her physically desirable. "Tell me what a man finds sexually attractive and I will tell you his entire philosophy of life," says one hero to another. "There is no conflict between the standards of his mind and the desires of his body."

In 1954, when the huge novel seemed to be nearing completion, Rand and Blumenthal—heroically rechristened Nathaniel Branden—summoned their spouses together to announce that they had fallen in love. This meeting and its trail of repercussions were first disclosed in Barbara Branden's not unsympathetic 1986 biography, *The Passion of Ayn Rand,* and were fully elaborated upon in *Judgment Day,* Nathaniel's own memoir of the great affair, published in 1989—over Rand's dead body, as her supporters have been quick to note. At the time, however, it was apparently Rand who did all the explaining: "By the total logic of who we were—by the total logic of what love and sex mean—we *had* to love each other." The pair did not expect this development to affect their marriages. As Rand put it, "It's right and rational that a sexual affair between us can last only a few years. I could never be an old woman pursuing a younger man." She was nearing her fiftieth birthday. Had they been lesser people, the four of them could not have accepted such an arrangement, she explained, "but we're not lesser people."

When, in *Atlas Shrugged,* Dagny finally crashes into Galt, her longtime lover graciously steps aside in recognition of the better man. Frank O'Connor, however, took to drink. Rand hadn't the time, or perhaps the capacity, to notice. It took her two years to write the ideological climax of her book, a sixty-page speech in which Galt details the strikers' demands in terms of the moral bases of capitalism and the universe. When she finished the book, publishers begged her to cut the speech—to cut anything. "Would you cut the Bible?" was her reply. At Random House, Bennett Cerf took the book whole. In 1957 *Atlas Shrugged* appeared, dedicated jointly to the author's husband and Nathaniel Branden.

The reviews were a massacre, and this time the attacks came harder from the right than from the left. Conservatives were appalled at the prospect of being identified with a popular novel that drew so many of their positions to their wildest conclusions. The harshest charges came from William Buckley's *National Review*, in which Whittaker Chambers, no less, covered all the odds by comparing Rand in her godless heresies to Marx, and by vouching that the implicit message of almost every page she wrote was "To a gas chamber—go!" Added to the accusation of literary fascism was the perhaps more disturbing charge, to the author, of not being literary at all. A few general magazines praised her storytelling and her "eloquent flow of ideas," but the critics who mattered had turned from the political criteria with which she had felt herself on fighting terms, at least, to a credo of language and style by which she simply did not exist—no matter how many readers among the once revered masses thought otherwise.

After the reviews, she cried every day. She asked Branden not to tell the others in their group. "Galt would handle all this differently," she confided to him. "I would hate for him to see me like this. I would feel unworthy, as if I had let him down." She spent most of her days at her desk rereading her book. The affair with Branden slackened and left off, and he took over the management of their lives. Within a year, he was successfully marketing her ideas, packaged under the brand-name Objectivism and disseminated nationally through lectures offered by the Nathaniel Branden Institute. Rand would sometimes appear to answer questions after the lectures, but she was so irascible in her replies that she was eventually discouraged from attending at all.

Rand's combative style served her well on college campuses, though, and she herself became a highly popular lecturer: the Camille Paglia of the early sixties, railing against "intellectual hoodlums who pose as professors." Appearing in her signature black cape and a dress pinned with a gold dollar sign, she easily converted to atheism and individualism all those whom J. D. Salinger had not already converted to Buddhism. Who is Howard Roark, after all, but Holden Caulfield with a heavy dose of testosterone—the unique young soul of integrity moving detached through a world of phonies? By 1962 English professors were exchanging nervous bulletins about Rand's possible effect as "the single writer who engages the loyalty of the students." Out in the real world, meanwhile, Gore Vidal noted with some alarm that "in my campaign for the House she was the one writer people knew and talked about." He pronounced her so-called philosophy "nearly perfect in its immorality."

How to account for the rampant success of a godless immoralist in the heartland of Christian democracy? Was Rand what her critics said she was? Was she what *she* said she was? A reader can hardly get through more than a page or two without sniffing the burning fuel of subverted emotion, or seeing political outrage as a mere component of her recoil from the broader offenses of mankind (and especially womankind) upon her senses: dirt, sweat, fat, sagging breasts, softness, confusion, ill-fitting clothes, ugliness—all endanger the heroic ideal. Dominique would rather see every one of Roark's buildings explode than see their balconies hung with diapers. (Rand had no children and her work stops just short of proving they don't exist.)

Still, *The Fountainhead* was not *Mein Kampf*, as one reviewer had claimed it to be, and American readers were not following Rand into fascism. A little bravura selfishness has always seemed attractive in our heroes and heroines, if only to lighten our perpetual burden of doing good. And, in fact, Rand's novels are riddled with syntactical loopholes that permit, in bluffest disguise, just the compassionate behavior she claims to disavow. Defending an outcast student, one heroine carefully explains, "I am not doing this *for* her, I am doing it *against* twenty-eight other girls"; another offers consolation to a wounded soul "not because you suffer, but because you haven't deserved to suffer." In a host of neatly adjusted proofs showing that "selfishness" will produce the same results as charity, only more efficiently and with wider benefits, Rand reveals herself to be a closet Samaritan. Sidney Hook, the only philosopher of repute to glance in Rand's direction in her years of highest popularity, pronounced her a "paper tigress" and a principled, if confused, opponent of totalitarianism. The last letter Rand wrote before she died contained a gift of money to her niece for a big Christmas dinner. She was being selfish; she herself couldn't bear to know that the woman was depressed.

But, whatever Rand stood accused of, her books would not stop selling.

Slowly she began to live again. She began writing essays for the *Objectivist Newsletter,* which she published with Branden, and she later collected these articles under such cheerfully antagonistic titles as *The Virtue of Selfishness* and *Capitalism: The Unknown Ideal.* She broke with Random House over an essay titled "The Fascist New Frontier"; Cerf drew the line at "any book that claimed Hitler and Jack Kennedy were alike." She wrote a newspaper column on current events; she took her first airplane ride; she accepted an honorary doctorate of letters. And she told Branden she was ready to recommence their love affair.

That was 1963, two years after Branden had fallen in love with a twenty-one-year-old fashion model who sat in the front of the class through his weekly lectures on rationalism and unity. ("The expression on her face was rapture," he recalls.) He put Rand off. He was divorcing his wife, and he was concerned for Rand's husband. Rand became suspicious and, as he records, began to ask with renewed insistence, "Am I still a sexual being to you?" After all, she was still his sworn and patented Goddess of Objectivity. He could only say yes. It took him five years, until July 1968, to admit his panicky thoughts even to himself. In the meantime, he began to have some doubts about her philosophy.

When Branden finally broke the news, he was four years into an affair with his fashion model, whom Rand described as "the physical type of my heroines." Rand knew her as an attendant in their circle, sufficiently devoted to have adopted the last name of a *Fountainhead* character—she called herself Patrecia Wynand—when she set out on an acting career. At first Branden told Rand only of the difficulties that had grown with the contrast in their ages, at sixty-two and thirty-eight; she pronounced him "a scoundrel." He then confessed his real attachment, and she went into a fury—"The girl is nothing!"—and gave him two months "to regain his mind." She demanded a statement "that it is his intellectual judgment that my position on Patrecia is understood by him and is objective." When he refused, she cut him out of their business enterprises, removed his name from the dedication of *Atlas Shrugged,* sent a letter denouncing him for unspecified moral transgressions to all Objectivist subscribers and followers, and—he claims—tried to stop the publication of his masterwork, *The Psychology of Self-Esteem.* Barbara Branden recalls Rand wearily confiding, "He has forced me into a permanent ivory tower. He took away this earth."

Branden married Patrecia and moved to California, where he became a psychotherapist and the author of such studies as *The Psychology of Romantic Love* and *If You Could Hear What I Cannot Say.* Rand, turning to her husband, found that he had become a shell—a man in his seventies troubled with alcoholism, a failing memory, and an anger incomprehensible to them both. In her outward life there was little change, except that her targets proliferated and her expression became more virulent as the sixties turned into the Sixties and left her behind. She was overwhelmed by the political tides of youth moving to the left, but she went down fighting. She was against the Peace Corps, against the draft, against the war in Vietnam, against the student movement, against the universities, against the feminists, against the religious right, and, eventually, against the Libertarians who claimed her as their guiding spirit. Finally, turning against her friends for their flawed tastes in painting or music (she was devoted to Dalí and Rachmaninoff), she immersed herself in looking after her husband.

In 1974 Rand was diagnosed with lung cancer, but she recovered well from surgery. The following year, announcing her plans to discontinue what was now called *The Ayn Rand Letter,* she wrote that the evils destroying civilization had become too "smutty and small" for her to wish to address. "Perhaps the last cultural fad one could still argue against was Karl Marx," she wrote. "But Freud?" Society had descended beneath reasoned argument. She continued to air her immutable philosophy on television talk shows, though, declaring the joys of her life to have been "achievement and romantic love, my husband." She responded to the perpetual questions about nearing death unarmed by faith in God by saying that she was not afraid of dying, only of being left behind. "The worst thing is to lose someone you love." After Frank O'Connor's death in 1979, she cited as proof of her lack of belief in a hereafter the fact that she had not committed suicide in order to join him.

Rand died in March 1982 and was buried in Valhalla, New York, with her husband's photograph on her chest; Kipling's "If" was the only graveside reading. In New York, hundreds of people—well-known economists (among them the ever-loyal Alan Greenspan), math and philosophy professors, kids in jeans—gathered at the funeral home, where she was laid out before a six-foot dollar sign. Guards were reported ready to turn Branden away should he try to attend, but he did not try.

All her books were in print and widely read. Then, more than a decade later, an important new title was added: Мораль Индивидуализма, which translates as "The Morality of Individualism," is a compilation of extracts from Rand's essays and novels, published in 1993 in the former Soviet Union. The first edition, of five thousand copies, sold out in Moscow in two days. This slim volume contains some of Galt's big speech from *Atlas Shrugged,* but not the book's Marxian fable of the decline of the Twentieth Century Motor Company—which concludes, in any case, just as everyone by this time would have guessed. The outcome had seemed less obvious forty years before, when she wrote, "It ended the only way it could end: in bankruptcy."

And how is the prophet honored? In 1993 it was reported in the land of the dollar that Michael Milken received twenty-two gift copies of *Atlas Shrugged* while in prison. Libertarians have propelled many of her most cherished antigovernment ideas into the forefront of American politics. And then there is Camille Paglia's brightly belligerent spirit and her Victoria's Secret Nietzscheanism—seeming direct from Dagny Taggart, complete with the sexual-industrial thrill of crossing a bridge ("I think: *men* have done this. Construction is a sublime male poetry") and of economic machismo ("The capitalist distribution network . . . is one of the greatest male accomplishments in the history of culture"). For the truly dedicated, there are institutes and mailings and conferences with such titles as "Ideas for the Rational Mind," offering lectures by Rand's closest followers and mini-courses in such subjects as "American Business Heroes" and "The Four Giants of Philosophy"—Plato, Aristotle, Kant, and Rand.

Still, her achievement wasn't what she had envisioned, which was a novel with the force of, say, *Das Kapital.* Or, more to the point, a novel like Nikolay Chernyshevsky's *What Is to Be Done?*—the Russian *Uncle Tom's Cabin,* which converted Lenin to revolutionary socialism at the age of fourteen. Even this would have been only a means, however—an intermediate step. On a deeper level, Rand claimed, she had never wanted to write about politics at all. That was merely what she had had to do in order to help secure a world in which she could write what she really liked: romances, adventures, stories that would be like the American movies she had seen in Petrograd—exuberant stories of what she called "the benevolent universe."

Around the time of the break with Branden, she finally began such a story—a novel about a ballerina who falls madly in love with a man who doesn't want her, and whose rejection causes her terrible suffering. Rand was not able to carry out much of the project, and returned her publisher's advance when she became ill. But she knew from the start what the story's ending would be, and she wrote some of it down. The beloved man finally awakens to the qualities of the woman who has worshipped him with such passionate intensity, and returns her love. He begs her to forgive him for all the pain he has caused her. The last line of Rand's only work to be liberated from politics is the woman's simple and consoling reply: " 'What pain?' she asked."

Memoirs of a Revolutionary

Doris Lessing

On a blanket on the lawn under a cedrillatoona tree, in Southern Rhodesia, in 1942, a young woman sat explaining to her children why she was leaving them. The boy was three years old and the girl not yet two, but she believed that they would understand someday and thank her. She was going off to forge a better world than the one she herself had grown up in, a world without race hatred or injustice, a world full of marvellous, unselfish people. She later admitted that she really hadn't been much interested in politics; what had drawn her to the local Communist Party was the love of literature she'd found she shared with its members—some ten to twenty mostly well-off, young white people who'd formed a self-designated Party branch in nearby Salisbury—and the sense of heroic expectation that surrounded their lives. They were eagerly awaiting the revolution they believed would end the hateful white regime, and if their activities were principally confined to meetings and disputes and love affairs with one another, their goals were undeniably noble. The newest comrade had not yet begun to consider the disparity between their professed goals and their actions, possibly because she could not afford to consider the disparity in her own.

Doris Lessing's stingingly self-mocking account of her escape from maternal to global responsibility appeared only a few years ago, in the first volume of her autobiography, *Under My Skin.* Her thirty-sixth published book, it brought her saga only up to the point of her departure from Africa, far short of her having earned "an identity," as she had put it when she'd thought of facing her children again, that would "justify my having left them." Lessing's second volume of memoirs, *Walking in the Shade,* carried her story to its literary climax, in 1962, with the publication of her most celebrated novel, *The Golden Notebook*—continuously in print, widely taught, and by now an emblem of the social history it once helped to shape. Many of the scenes in both volumes of memoirs are already familiar, since Lessing has drawn on the particulars of her experience throughout the broad reaches of her fiction, characteristically joining the daily realities of life on the veldt or in a London flat to the biggest political issues of the age.

"After all, you aren't someone who writes little novels about the emotions. You write about what's real," a Party comrade assures the emphatically autobiographical heroine of *The Golden Notebook,* Anna Wulf. A fully self-conscious specimen of the "position of women in our time," Anna is also an individual of high neurotic distinction, the very model of the modern madwoman who has found her way down from the attic to the bedroom only to stand fumbling for decades with the next set of keys. It is Anna—an earnest revolutionary with a weakness for scarred and brooding men—who patiently writes and assembles the many parts of *The Golden Notebook* itself: an old-fashioned baggy monster of a novel, at once didactic and feverishly intense, thick with sermons and stories about African racial policy and Soviet Communism and the modern male's inability to love and the vaginal orgasm and the question of whether women can ever be free. Anna's goal is to write a book that will change the way people see the world, and many readers of *The Golden Notebook* claim that Lessing herself accomplished something remarkably like that. Many others, however, have viewed the novel as simply the era's fiercest declaration of what its author plainly termed "the sex war," an old subject newly named in accord with precipitously heightened stakes and lowered standards of attack. It is perhaps only with the publication of Lessing's memoirs that we have been offered a portrait of the artist as a twentieth-century woman that is as dark and turbulent as that of Lessing's infamously semiliberated heroine, a portrait as disturbing as it was evidently meant to be.

▪ ▪ ▪

"If marxism means anything," Anna rebukes the comrade who has complimented her on her writing's worldly scope, "it means that a little novel about the emotions should reflect 'what's real' since the emotions are a function and a product of society." It was Lessing's insistence on this kind of connection that earned her a fervent readership during the 1960s, when the slogan "The personal is political" seemed a newfound formula of escape from the cultural lockstep of the fifties. But long before she'd encountered the idea in Marxism, Lessing had absorbed such a connection through the novelists she'd cherished in her youth: principally Olive Schreiner, Virginia Woolf, and D. H. Lawrence, all of whom advocated social regeneration through a renewal of the individual which variously mingled sex and spirit. These writers had provided Lessing with a saving vision, not for the future of the world but for her own past. "The personal is political" was more or less the formula she had struck on in her attempt to forgive her parents and explain her childhood.

For when viewed in the proper historical light, as she came to see it, the girl born Doris Mae Tayler was not merely the daughter of a crippled English soldier and the military nurse who'd married him, out of exhaustion and pity, after her true love's ship was sunk by a torpedo; she was the daughter of the First World War, the howling product of shell shock wedded to disguised defeat. And the reason that her birth had been the unwelcome disaster that, year after year, her mother had recounted to her in all its horror—forceps, agony, bitter disappointment—was not that her mother actually hated her but that the world had long ago cut out her mother's heart.

"My God, in what a century have you caused me to live!" is Lessing's epigraph for her "Martha Quest" series of autobiographical novels; the words are attributed to St. Polycarp, who wrote them down in A.D. 156. Not even sharing the historical joke across millennia, though, not even this vast perspective, could assuage the particular pain of having been born in October 1919, with the Great War barely over and her parents married exactly nine months and so thoroughly unfitted for their roles that it was the doctor who suggested that she be named Doris, since neither the new mother nor the father could think beyond the one name they had prepared—which was Peter John, after the boy heroes of *Peter Pan.* (Mrs. Emily Maude Tayler called her husband Michael—his actual name was Alfred—after the younger brother in the Barrie story; she herself was called Maude, in tribute to Tennyson.)

This difficult birth took place in rural Persia, where, after the war, Alfred Tayler had got a job in a bank. Having lost a leg and all his political idealism somewhere near Passchendaele, the thought of ever living in England again filled him with a ranting fury. His wife, however, carried England along with them as unconditionally as she could. For the first year of their daughter's life she was continuously starved—later she was told how she had screamed night and day—because the strict English feeding schedule Mrs. Tayler observed didn't take into account the thinner milk of scrawny Persian cows. This was a story her mother repeated to the girl with relish, as though it were a tremendous joke they both might share. (Recalling such stories decades later, Lessing would wonder if her mother, whose own mother had died when she was three, had simply not known *how* to behave like a mother, never having been shown.) But the big white nursery in Kermanshah that shimmers like a fever dream in Lessing's novels seems to have been a place of many varieties of deprivation. She was hardly more than a baby herself before she was sharing the room with a little brother who did not scream or protest or cause any trouble, and who—either as a cause of this behavior or as its result—was genuinely loved.

After her father refused a job transfer to a larger city, the family returned to England for a visit, crossing by train through Russia—Baku to Moscow—in the spring of 1924. Lessing believes they were the first foreign family to be allowed to travel normally after the Revolution, although all that impressed her at the time were the filthy platforms thronged with desperate beggars crushed up against their windows—especially terrifying were the orphaned and abandoned children—and her anguish when once the train pulled out while her mother was still off searching for food. Her father, with his useless stump of a leg, was as helpless in the circumstances as the two children, and in the days that it took Mrs. Tayler to catch up with her family—a near miraculous feat, considering that she was alone and spoke no Russian—their daughter swiftly came to understand how pitiably dependent they all were upon this proper Englishwoman's unfathomable resourcefulness and strength.

A few cold and miserable months after their return to London, Alfred Tayler went to visit the 1924 British Empire Exhibition at Wembley. Passing the posters at the Southern Rhodesia stand, he decided that a fortune could be made in that country growing maize; the family shipped off for Africa a few weeks later. Lessing's strongest memories of the voyage were of being tossed overboard as a mark of jubilation at crossing the Equator (she was fished out half-drowned and sputteringly enraged) and, even worse, of being locked in the family's cabin while her parents went off to parties that she longed to attend. One night, in her fury at being left behind, she hacked up one of her mother's evening dresses with a pair of nail scissors. She was five years old, and she recalls that these goading torments already felt familiar.

Her mother had packed many more evening dresses along with the books and pictures and heavy silver that were sent in trunks to follow behind the three adults—they had brought a governess—and the two children who, upon arrival in Cape Town, were disposed in a covered wagon that was set to sixteen oxen and drawn for five days until it reached the patch of virgin bush that Alfred Tayler had bought: fifteen hundred acres of wilderness a few hundred miles south of the Zambesi River. It was two or three miles to the nearest farm, and several more to the dusty crossing where a handful of merchants constituted a bleak approximation of a town. In between these scattered settlers there was nothing. Resident African blacks had been driven onto reservations decades before, when Cecil Rhodes had annexed what had been Mashonaland for England, and renamed it for himself. It was now late 1924, and the maize market was going into a slump.

The family built a hut of whitewashed mud and thatch and dung, which was hung with Liberty curtains; the parlor piano arrived a few months later. They had put in about a year of stumping and clearing the land in an effort just to open some fields on which to begin to farm when Mrs. Tayler retired to her bed with a condition that her husband respectfully called "a heart." It was another year before she got up again. When at last she did, she cut off her long brown hair and made a ceremony of throwing it in the trash—her children wept—and then she set to work, trying to enforce a series of her husband's impracticable schemes in what became a lifelong plan for "getting off the farm." With every year they got poorer and more into debt. The hut that was meant to last until a real home was built accumulated new rooms and a porch and became the real home itself. Lessing writes that her mother—"a vibrating column of efficiency and ruthless energy" whom she often feared might simply mow her down—gave no sign of ever realizing that the man she had married was not a farmer but a dreamer. Instead, she told anyone who would listen that it was the day-in, day-out sacrifice of being a mother that had ruined her life.

"I was in nervous flight from her ever since I can remember anything, and from the age of fourteen I set myself obdurately against her in a kind of inner emigration from everything she represented," Lessing writes. She had been sent away at seven to the first of a series of boarding schools, occasionally frightening and always lonely places that made her long for the home she felt pushed out of; this early expulsion is one of the most grievous in Lessing's litany of childhood injuries. At fourteen she came home to recover from an eye infection and declared she was never going back. She continued her education entirely on her own, from whatever books she ferreted out herself, mostly novels and works of popular psychology: the Martha Quest series begins with the fifteen-year-old heroine lolling about the family porch, after having refused to return to school, trying to shock her mother with the carefully exposed cover of a book by Havelock Ellis. The heroine's age and the sex book are more than signs of the usual adolescent rebellion. This is the beginning of the central drama—one might say the central battle—of Lessing's life, for she was just discovering that she had, in sex, a powerful weapon to use against her mother.

From a stout little girl she had turned into an uncommonly good-looking young woman, and the early Martha Quest volumes are filled with the proudly luxuriant pleasures of her own flesh. Tanned, golden, freckled, firm, weighing a little more or a little less, the heroine focuses all her curiosity and desire on the fascinating object in the mirror. The men who take her to dances or fumble with her clothes or want to marry her are a largely indistinguishable lot; even in bed she does not really see them so much as she imagines what they must be seeing when they look at her. All this attention was, at least, a wondrous distraction from what was by then a firmly ingrained sense of personal injustice, but it was even more of a distraction from her announced goal of becoming a writer.

In 1937 Lessing moved to the city of Salisbury, about a day's trip from her parents' farm, and took a secretarial job to support her ambition. (Martha Quest would do exactly the same.) She wrote two apprentice novels, very much under the influence of Lawrence, and tore them up. She had gone back to working on short stories, as an exercise in technique, when she found herself suddenly married—at nineteen—and writing nothing at all. The passive tense is indicative of Lessing's own attitude to the way things happened: in her memoirs her husband seems less a person than an unavoidable minor stage in her evolution (one could imagine she married him just for the promise of his name: Frank Wisdom). In *A Proper Marriage,* the second volume of the Quest series, Martha's husband is the dullest man in the dullest city in the dullest country in the world, and seems no less externally inflicted on our heroine than the heat and the dust. Lessing's attitude is almost equally passive in regard to the birth of a son approximately nine months after her marriage, and then a daughter a year and a half after that. It wasn't that she was ever in love, she tells us, or that she had wanted to get married or to have children: it was all a matter of the "Zeitgeist." The war was coming, everyone was getting married and having babies, so why not she?

Lessing clearly intends this admission as further evidence of the power of History over her life, but in this instance her argument lacks force, and—with the "Zeitgeist" deemed a nearly personified force that bestows significance even as it removes responsibility—she seems set on an excuse rather than a philosophy. Given Lessing's determination to locate the cause for her actions outside the sphere of her free will, she makes a far more substantive case for an innocent young woman's betrayal by literature:

> "If you read novels and diaries, women didn't have these problems. Is it really conceivable that we should have turned into something quite different in the space of about fifty years? In the books, the young and idealistic girl gets married, has a baby—she at once turns into something quite different; and she is perfectly happy to spend her whole life bringing up children with a tedious husband. Natasha, for instance; she was content to be an old hen, fussing and dull; but supposing all the time she saw a picture of herself as she had been, and saw herself as what she had become and was miserable—what then? Because either that's the truth or there is a completely new kind of woman in the world, and surely that isn't possible, what do you think, Caroline?"

This quizzical speech is addressed by Martha Quest to the small daughter—aged roughly two—who stands in for Lessing's pair of children in the otherwise nearly stroke-by-stroke autobiography of *A Proper Marriage.* Martha is struggling to understand how, in contradiction to all she'd been promised, being a suburban wife and mother was boring her to extinction. Although Martha's most important life principle is described as "hatred for the tyranny of the family," she has entered marriage without a suspicion that this tyranny might prove as hateful in the family she is creating as in the one she escaped; even now, she doesn't realize that she is ready to make her break. Lessing was also slow to admit to herself what she was planning to do. In her memoir, she writes of the emotional withdrawal from her children—particularly from her tiny daughter, Jean—that preceded her departure, a withdrawal that she only dimly understood was making it possible for her to go.

At first it seems merely odd that the ensuing familial drama of abandonment, outrage, and emotional appeasement—in both novel and memoir—is played out most passionately not between Lessing and her husband or her children or the inevitable, temporary "other man" but between Lessing and her mother. And then this comes to seem the central point, and to offer what may be a more cogent explanation for Lessing's avowedly random, victim-of-History behavior. Lessing often complains of having been unable to argue openly with her mother, who broke into childish tears at any sign of rebuke; disarmed by guilt, the good daughter's only means of defense were counterattacks so carefully disguised as to be inexplicable, in content and vehemence, even to herself. To a reader of the accounts of this continuous subliminal warfare, there comes the awful but unavoidable thought that Lessing may finally have duplicated her mother's perfect two-child family precisely in order to throw it away—to demonstrate once and for all what she thought of her mother's sacrifice and the price it was meant to exact from her own life in return.

When she first moved out to start reshaping the world, she went only as far as another neighborhood in Salisbury. Still, she thought it best to break entirely with her children and she did not see them—except by her mother's occasional design, at events she endured as the punishment she knew they were meant to be. She remarried as soon as her divorce came through, in 1943, and a high point in the drama appears to have been reached when she announced that she was pregnant, at Christmas of 1945. Even her generally all-accepting father, she writes, came to her to complain: "Why leave two babies and then have another?" Her mother, she reports, was "fiercely, miserably accusing." As for Lessing, even half a century later she reasons that her actions and desires—for she specifies that she truly desired this baby—were due to the uncontrollable effects of wartime: "Just as in 1939, when I got pregnant with John, and then, so soon, with Jean," she writes in *Under My Skin,* "I believe it was Mother Nature making up for the millions of dead." Who was she to deny the most powerful of mothers?

▪ ▪ ▪

The bourgeois family will vanish as a matter of course . . . with the vanishing of capital.

—KARL MARX AND FRIEDRICH ENGELS,
The Communist Manifesto

It was her second marriage that turned the young woman who was born Doris Tayler and became Doris Wisdom into the writer Doris Lessing. Her new husband, Gottfried Lessing, was a one-quarter-Jewish German refugee of wealthy Berlin origins who'd fled to England in 1937, and who'd become a fervent Communist while living on the skids in London. Shipped off by the English to the wretched little colony of Southern Rhodesia, he cut a notably glamorous figure: serious, handsome (Lessing writes that he looked like Conrad Veidt), and, while still in his twenties, possessed of a worldly sophistication hardly known to the local population. He soon became the leader of the tiny group of amateur Communists—Moscow never acknowledged their existence—who provided the committees and meetings and arguments that would prove so potent a force against the boredom of young motherhood, if against nothing else.

It was the cause she loved—or, at least, it was the excitement of having a cause—and not the man. For on closer inspection, his ex-wife confides repeatedly, Gottfried Lessing turned out to be stiff and repressed and humorless and terrible in bed. (One of the few truly comic, almost Mary McCarthyesque lines in Lessing's memoirs is her recollection that her orthodox-Marxist anti-Freudian husband "once refused to speak to me for a week because I joked about the unconscious, and it was not even his.") Of course, she had discovered all these defects long before they married. The Lessings' official union was more or less a political act (the category is complicated by the fact that politics was so much more romantically charged for them than sex). He was an enemy alien, and she feared that his affair with a woman who "had so recently been unpleasantly divorced" threatened to bring dangerous attention to his Communist activities. From this and almost all other dangers, British citizenship would save him. Ergo: "It was my revolutionary duty to marry him."

From the beginning they had vaguely intended a divorce, but they had to delay proceedings until they received news of Gottfried's admittance to the rights of the Empire he despised. During these months, he apparently began to harbor some hope of developing a real marriage while she had several (later much detailed) affairs, and somehow they agreed they might as well go ahead and have a baby. The Communist group thought this a wonderful idea; it would be their first. The Lessings' son, Peter—the name that had once been reserved for his mother—was born in October 1946 in the maternity hospital where his mother had given birth twice before. "I was in love with this baby," Lessing writes, making an implicit distinction that is difficult to read without wincing. Equally difficult to read is her account of the complex family picnics engineered by her mother, at which her own two older children got to view their otherwise absent mother with a new baby in her arms. Grimly, Lessing jokes—she is joking, isn't she?—"I have had doubts about the joys of the extended family ever since."

But she had no doubts anymore about what she wanted her future to be. She had begun writing again, laboring on an enormous Balzacian comedy of colonial life, which she eventually cut by two thirds on the orders of a London publisher—here begins a famously ambivalent relationship to publishers—and transformed (rather to her own surprise) into a heated sexual tragedy. *The Grass Is Singing,* Lessing's first published novel, is an "African" work in both its setting and its theme, which is the neurotic and often erotic relations of whites to blacks. Yet beneath its skin, the book concerns the same essential subjects that would make Lessing famous more than a decade later, when the times caught up with her: a woman's diminishment in marriage, the destructive power of sex, and madness as a reasonable result of a maddening society.

Set on an isolated South African farm, the novel's central figure is a desperately lonely and repressed woman whose marriage to a kind but weak man leads her to catastrophe. Mary Turner is the first in a long line of Lessing heroines to ruin her own substantial health and strength in trying to create a man—or even the illusion of a man—who is stronger than she is. And she is the first to try to run away, back to the self-sufficiency of her life before marriage, only to discover that the world has passed her by and she is no longer young or brave enough to begin again. (Mary McCarthy's contemporary story "The Weeds" places the same sort of marital disaster in a very different social ambient; McCarthy's scene of a sadly mystified, much hated husband arriving to bring his shattered wife home is almost emotionally identical with Lessing's.) Race comes to the fore when Mary Turner's near hysterical loathing of her husband's black workers crystallizes into a self-annihilating attraction to a virile "houseboy," with whom she develops a tortured but never sexually overt relationship. In her memoirs Lessing reports that Blanche Knopf had offered to buy the book if Lessing put in an "explicit rape," which the publisher rationalized as being "in accordance with the mores of the country." (Considering American best-seller lists, the question is, of course, which country?)

Continuously observed through the heroine's eyes, the African servant appears as little more than a standard embodiment of "dark" masculine force. But Lessing adds a strange twist to the old cliché—even beyond the obvious subjectivity of the heroine's impressions—when the mentally disintegrating woman begins to confuse the man, in his sexually menacing presence, with her father. And, suddenly, one realizes that this is not primarily a racial story after all. The woman's association with her father is evoked by a nightmare in which she relives the times when, as a little girl, he would press her head down into his lap while she was meant to be covering her eyes for hide-and-seek, and she had nearly suffocated from the "unwashed masculine smell" and the fear. Even as a dream, this episode—the only mention of the heroine's childhood—reads like an intrusion from a different book; in Lessing's memoirs she reveals it to be autobiographical, a part of the insidiously sexual roughhousing her father inflicted on her throughout her early youth. It is one of several such episodes that recur in her work over the years, with the depictions growing ever more graphic and angry.

Alfred Tayler was grievously ill during the period when Lessing was writing and revising this novel; her mother was once again his dutiful nurse. Lessing records that on the day her father died she was summoned to his bedside, but that she chose to remain at home and bathe her baby son instead. She didn't believe that he would die, she writes, and in any case she did not want to be there when he did, even though she knew his real death had happened long ago, during the First World War. Besides, what she describes as her impulse to pull out her hair or rake her cheeks with her nails would not have been tolerated at Salisbury Hospital; so she stayed at home and went on bathing her baby. If it was only after her father's death that Lessing was able to leave Africa, the reason seems to lie not in release or in sorrow, however, but in fear: her mother's formidable energies were once again unleashed. It was 1949 when Lessing packed up the manuscript of *The Grass Is Singing* and two-and-a-half-year-old Peter and went off to Cape Town, where at long last she boarded a ship to London.

"I was free. I could at last be wholly myself," she exulted as she caught sight of the towers of the literary promised land. She was thirty years old and a certified writer. "A clean slate, a new page—everything still to come." Any expectations that Lessing's tale might soar into a triumphant resolution to *A Doll's House,* however, are quickly smashed. Looking back, Lessing is sardonic, even accusing, as she marvels at what she terms her adolescent sense of self-possession. In the modern self-experiment of a life she had begun, freedom would turn out to be much harder to use than it was to win.

Her account of her early London days is a catalogue of the hardships of single motherhood, what with getting up at five, working for too little money during the hours when her son was at nursery school, feeling ever duty-bound and exhausted. Her time for writing came at night, when Peter had fallen asleep, although she was often so tired she simply fell asleep herself. But worst of all was the tremendous loneliness and exclusion from adult society. London was still a dark and blinkered city, a barely recovered war casualty itself. She describes walking the empty streets alone at night, tantalized by the lights and the glimpses of fellowship in the pubs she sometimes passed. And one feels that if only these pubs had been a little less beery or more welcoming to a foreign woman, Lessing might not have committed "probably the most neurotic act of my life," about a year after her arrival, and officially joined the Communist Party.

> I did know it was a neurotic decision, for it was characterised by that dragging helpless feeling, as if I had been drugged or hypnotized—like getting married the first time because the war drums were beating, or having babies when I had decided not to—pulled by the nose like a fish on the line.

Ironically, it was just then, in 1950 in London, that a group of American and European ex-Communists—among them Arthur Koestler, Richard Wright, and Stephen Spender—published their personal testimonies in the great confessional account of intellectual disillusionment with Communism, *The God That Failed.* Had Lessing read even Richard Crossman's introduction to the book, she would have found words for what she was to spend much of the next several years learning to admit: that "despair and loneliness" were principal motives for conversion to the Party faith, and that its lasting attraction for many a wandering soul existed not in spite of the sacrifices it required but precisely *because* the Party "demanded everything, including the surrender of spiritual freedom."

Crossman also notes that, despite their long disaffection, none of the volume's contributors deserted the Party "willingly or with a clear conscience." The same is certainly true for Lessing, who devotes much of her memoir to anxiously explaining why she stayed on past so many protests of her conscience. Riddled by doubts from the start, she was increasingly dismayed by the petty corruption of the British Communist bureaucracy and by evidence of the deadly Soviet variety, impossible for even the most faithful to ignore after Khrushchev's 1956 revelations of Stalin's atrocities and the invasion of Hungary later that year. In fact, she was sufficiently outraged about Hungary to write a "passionate letter protesting it" to—astonishingly—the Union of Soviet Writers. Perhaps no one has taken the power of the pen more literally; one longs to have witnessed the bleak Gogolian farce out of which a "conciliatory letter" was produced for her edification and sent by return post. And still she struggled with turning in her Party card. Once again, she blames the undertow of history itself; if her feelings were neurotic, her thinking belonged to the "Zeitgeist." What she was always secretly hoping for, she writes, was the emergence of a few pure Russian souls to put the system back on its true path.

The most perplexing and dispiriting of her adventures, however, occurred when she came upon just such a soul, and the message was not what she wanted to hear. In 1952 she was visiting the Soviet Union and touring a collective farm when a classic Tolstoyan old farmer in white stepped forward to cry out that everything the foreigners were being shown was false, that life under Communism was terrible, and that they must go back to Britain and tell the truth. Lessing now calls this the bravest act she ever witnessed, since the old man had to know that "he would be arrested and disposed of." It is unclear how much she herself knew or was willing to acknowledge at the time. She reports that after a banquet and a viewing of "presents to Stalin from his grateful subjects," she went to sit outside alone and brood. And it was apparently her shame over her aesthetic revulsion at those hideous presents—mostly boxes and rugs featuring Stalin's carved or woven face—that led her to the macabre conclusion of this episode: her decision to write a story "according to the Communist formula."

Stubborn as she obviously was, it did not take Lessing long to realize that this story—written with "greater simplicity" and "simple judgments of right and wrong"—was a failure. In fact, it would have been difficult for a reader to perceive her Communist sympathies during her Party-line years, although her political allegiances shaped much of her early work and helped to establish her reputation. *The Grass Is Singing,* which appeared in

1950, was widely applauded for giving British readers a rare view of how blacks were actually treated by the often idealized masters of southern Africa. (Only Lessing's Communist colleagues raised objections, complaining that the book's message had been "poisoned by Freud.") A volume of stories published the next year, called *This Is the Old Chief's Country*, ventured on the even more rarely addressed subjects of the cultural integrity of African tribes and the brutal process by which their land was stolen. But these stories are not diatribes; the politics are rooted in the lives of people, and the literary power derives from Lessing's explorations (often poisonously Freudian) of private conscience. White children slowly open their eyes to what their elders expect them to accept; farmers' wives seem to go mad almost as part of the job description, what with the loneliness and the heat and the necessary lies. Everywhere, sexual and familial dramas run like a scarlet ribbon through the black-and-white issues of African life, and bind the reader to them all the closer.

It is as a stylist that Lessing may be most nearly accused of being a Marxist. Her writing is plainly purposeful, and for a "serious" or "modern" author there are remarkably few occasions where a reader is caught up by the beauty or ingenuity of her words. Lessing explains this bluntness as a deliberate literary choice—based not on Marxist principles of simplicity versus decadence, however, but on her own idea of masculine versus feminine. She tells of her early, deliberate rejection of "the kind of writing usually described as 'feminine' ": that is, of a style "intense, careful, self-conscious, mannered." As a creative antidote, she forced herself to write a story she proudly titled "The Pig"—it begins, "The farmer paid his labourers on a Saturday evening, when the sun went down"—in a style she identifies as "straight, broad, direct," and therefore "the highway to the kind of writing that has the freedom to develop as it likes."

Lessing doesn't appear to see any irony in having developed such a proudly "unfeminine" style during the cramped night hours when she was able to write at all only by managing to stay awake later than her son. But then, it is remarkable how little these conditions affected either her ambitions or her output. *Martha Quest*, published in 1952, was the first of five planned novels in a series titled *Children of Violence*. The following year, Lessing published a book of novellas, followed immediately by Quest number two, *A Proper Marriage*. In accord with the series' several tributes to the work of Olive Schreiner, Martha's quest—destined to become political, spiritual, apocalyptic—originates in the heroine's personal defiance on a host of womanly matters, ranging from "the mechanics of birth control" and the legality of abortion to childbirth and the bonds of motherhood. Her toughest battles, however, are against herself. The scene in which Martha gives birth to her daughter is recounted in six brutally harsh pages of nonstop close-watched physical pain, during which the extravagantly self-critical heroine curses herself as a liar and a coward for calling out to a God in whom she doesn't believe, and then—even more infuriating to her, and a sure sign that the bottommost rung of Hell has been reached—for groaning "Mother, Mother, Mother!"

Lessing stopped writing her history to catch up on living it after *A Proper Marriage,* and the living was not easy. In 1957, with her sympathies still far from disengaged, she resigned from the Communist Party. Shortly afterward she sent her son away to school because she was afraid, as she says, that she had begun to need him too much for his own good. In *A Ripple from the Storm* (1958), Martha Quest realizes that her deepest emotion is a painful sense of having been always "excluded from some warmth, some good" that others naturally possess, and Martha's creator—for whom the memory of being left out of a shipboard party at the age of five was sharp as a knife almost seventy years later—now found herself more than ever lost in her search for a paradise she believed that she alone had never known.

Insisting on one's belief in the face of all evidence; seeing and hearing what you need to be true instead of what, horribly, is; hanging on to what you know you cannot live without. And then letting it go. Lessing's record of these long dark years after she left the Party is about the admission of colossal, sickening error and defeat. "All around me," she recalls, "people's hearts were breaking, they were having breakdowns, they were suffering religious conversions" over the collapse of Communism as a moral force. She points almost gratefully to one of the century's awful paradoxes: that, while the politically negligent helped make Hitler possible, it was "the most sensitive, compassionate, socially concerned people" who did the same for Stalin. This is not offered as a defense, exactly, but as evidence of a kind of mass delusion, in which the only imperative was to believe in something larger than the single self.

"Losing faith in communism is exactly paralleled by people in love who cannot let their dream of love go," Lessing writes. For her the comparison is not abstract. She suffered two disastrous love affairs during the years when she was wrenching herself away from the faith, long affairs that were more marriages than her marriages had been, and that left her feeling misused, empty, and nearly out of her mind with misery. In fact, she *was* misused, as she reports it both in her memoir and—in far more excruciating detail—in *The Golden Notebook.* Clearly, she was lonely and in need and had read too much D. H. Lawrence. But she was also a modern woman: she wasn't going to ask for anything that would make her seem wanting or weak. And the very modern "men-babies" that the age was producing certainly weren't going to give anything—there were now spoken rules about this—even as they pillaged her emotional store and absorbed all the loving and the cooking and the nursing and the sex that any sensible Victorian woman would have set at a far higher market price. No wonder she felt empty by the time they were sated and moved on. It was only decades later that she claimed she should have seen the danger signs, but that she had not wanted to.

"Sometimes I think we're in a sexual madhouse," remarks one of the shell-shocked women in *The Golden Notebook.* "My dear," a friend replies, "we've chosen to be free women and this is the price we pay." "Free Women" is, in fact, the title of a book that Anna Wulf begins writing, and it is meant to be ironic. Lessing's women are not only willfully blind in their romantic delusions but—and this *is* the price they pay—deprived of any further use of their corrupted and exhausted wills. Like Lessing blaming the "Zeitgeist" in her memoir, they have descended to the level where accepting pain is easier than taking responsibility.

What is most ironic of all, perhaps, is that *The Golden Notebook* has entered literary history as, in Lessing's words, "the Bible of the Women's Movement"—the novel that introduced the subject of women's liberation to American society. Lessing herself seems torn between laughter and tears at the thought. It is clearly her searching, shameless honesty that readers responded to so avidly; women had never talked like this in print before. But as a document of liberation, her book may be classed with Simone de Beauvoir's *The Second Sex* (which in 1949 announced the "free woman" as a type just being born), and perhaps with Richard Wright's *Native Son.* All three are works in which the angry subject is such a crippled specimen that hope and progress can be detected primarily in the sheer howling relief of the author's declaration: If I am a monster, it is because you have made me one.

At the time of their books' publication, neither Lessing nor de Beauvoir would have tolerated being called a feminist. De Beauvoir, the older by more than a decade, came to that self-definition only in the 1970s, while Lessing still finds the label irritatingly reductive. Certainly, both women pursued the lives of bold exceptions to their sex, and their writings betray the strain. Lessing has often been criticized for making that strain too much her subject; she complains too much. As for de Beauvoir, there was outrage just a few years ago when it was revealed that she had misrepresented the "free woman" aspects of her life with Sartre and other men, and concealed what amounts to a parallel existence of debilitating anxiety and chosen servitude; she misled us. But no reader of *The Second Sex* should have been surprised to learn that sexual freedom was not any easier for de Beauvoir than it was for Lessing. However high the goals each author envisioned for women in terms of work or art or liberation from domestic drudgery, in terms of relationships with men all that either could come up with was a repeated tale of emotional dependence leading to self-loathing, only leaving open the question of whether the sequence was not more explicable the other way around.

It seems an uncanny coincidence that de Beauvoir and Lessing were passionately involved with the same man, the famously macho Chicago novelist Nelson Algren, at the very time each was writing her big book. De Beauvoir nearly married Algren in the late 1940s, and *The Second Sex* (which Algren approved of and defended) clearly reflects some aspects of their relationship, which continued amicably until the end of de Beauvoir's life. Lessing's affair with Algren took place in the late 1950s—this was after her two barbaric "men-babies"—and he could hardly have approved of the appearance he makes in *The Golden Notebook.* Lessing was evidently angry enough to identify Algren by his rather distinctive first name, and then to describe him as a lifelong sexual cripple trapped in his terror and resentment of women, a man who used his well-honed charms to insinuate himself for the pleasure of hurting deeply, and whose essential nature she deciphers as "joy-in-destruction." But if "Nelson" is the worst of the men in *The Golden Notebook,* it is only because he does his damage so deliberately. The first-sex audience outcry over the piggishness and brutality of *all* the male characters in the novel—Harold Bloom still refers to Lessing's "crusade against male human beings"—prompted no more from Lessing than the deadpan reply that she really hadn't noticed.

In truth, Lessing is hardly any easier on women. Anna Wulf, blocked novelist, writes down her thoughts in four separate notebooks—black for

African experience, red for politics, blue for personal life, yellow for fiction—and longs for a sense of unity that will allow her to bring them all together: to write and live as a whole human being. It is a compelling theme, but, as so often with Lessing, the invented problem obscures a real one. Anna suffers not because her life is divided into different parts but because each part is rife with lies and self-delusions. Her true division—and Lessing's true division, as it is variously stated in almost every book she's written—is between the intellect and the emotions; or, between knowing what is right and doing the opposite anyway, because for Lessing the emotions always win. Anna's typical reason for walking straight into one more head-on male-female collision is: "I wasn't able to think at all, the emotional realities were too powerful. I think this is quite common with women."

Perhaps a more accurate title for *The Golden Notebook* would have been—to borrow Olive Schreiner's sadly unantiquated notion—"The Transitional Woman." Anna's conflicts are prompted by her attempts to live in a liberated present with a cast of mind and a nervous system still geared to the past. Lessing may have been bold enough to write about menstruation—many are the feminist gold stars she's been awarded for this so-called breakthrough—but what she expressed was shame and a monthly need for male comfort because of "the wound inside my body which I didn't choose to have"; Anna Wulf does more frantic blood-washing than Lady Macbeth. Lessing may have written about the female orgasm, but she believed that a woman could only have the "real" physical experience when she was in love. (In 1976 the Hite Report on Female Sexuality cited two MIT doctors' study of Lessing's physiological claims, concluding that the response she described was more akin to choking or sobbing than to sexual release.) And while Lessing's women may praise themselves for admitting—as men do not—that the physical and the emotional are related, this does not seem to work out as a female advantage. When a woman tries to overcome this essential unity of her nature—when she tries to act like a man, divorcing sex and emotion—it is only a sign that she is "morally exhausted," and heading for disaster.

What condition were women in to rally to such a philosophy and call it feminism? How did so many differently bred women—even those who had not had guilt-inducing parents or an affair with Nelson Algren—find their images in *The Golden Notebook*? Perhaps the most significant historical line in the book was Anna's startled observation, made after canvassing door-to-door for the Party before an election: "This country's full of women going mad all by themselves, all thinking 'There must be something wrong with me.' " By 1962, when the novel was published, the great postwar domestic myth was cracking top to bottom, but the housewives who were cracking first had barely begun to fathom how widespread their predicament was. Just a year later, in *The Feminine Mystique,* Betty Friedan chronicled the manipulation of feminine models and mores via American "women's magazines" that, from 1949, had steadily ruled out all female options save "Occupation: Housewife." According to Friedan, the existence of a terrible gap between the glossy pages and homebound reality was first suggested when an article planted by some bored editors in *McCall's,* in 1956, sold more copies than any issue in the magazine's history: it was called "The Mother Who Ran Away."

Lessing's distressed heroines tend to break down completely before they are able to put themselves back together—perhaps on the Marxist model that revolution must precede utopia. Lessing informs us that she herself managed to escape real breakdown through writing about it, but years of psychoanalysis may have helped. She first consulted a doctor in the mid-1950s, when her mother—"so pathetic, so lonely, so full of emotional blackmail"—came to London with the apparent hope of staying on as her daughter's secretary. Lessing went into treatment even before her mother arrived; just the letter announcing her intentions proved incapacitating. Lessing credits her psychiatrist, a Jungian who encouraged her to enter the minds of such unswayable role models as Electra, Antigone, and Medea, with giving her the strength to resist all appeals and to send her mother "home" to her brother's family in Africa. But there it happened that Lessing's brother and his wife—entirely without Jungian assistance—also managed to ward off the old woman's offer to devote herself to them and their children. In her memoirs, Lessing notes that her mother died not long afterward, at the age of seventy-three: "She could have lived another ten years, if anyone had needed her." As for herself, Lessing believed that she had been narrowly "saved."

Although psychiatry may have helped Lessing avoid the circumstances that would have led her to a breakdown, it has little bearing on her lifelong obsession with the subject. Doubtless Lessing's view of the nearly magical powers of madness is tied to the memory of her mother's having gone to bed for a devastating year and risen up hale and in command. Yet in her writing this kind of process is transformed into one more route to the desperately romantic goal of self-completion. When Anna Wulf "breaks down" she is accompanied by a lover in the same shattered state, a character Lessing calls Saul Green but who is identifiable as the American writer Clancy Sigal, already "pretty ill when he arrived" in London, according to Lessing's memoir. In *The Golden Notebook,* the furiously destructive couple actually " 'break down' into each other," as Lessing wrote in a 1971 introduction to the novel: "they hear each other's thoughts, recognize each other in themselves," and become so mutually attuned that they fill the pages of Anna's miraculously unified golden notebook together, so that "you can no longer distinguish between what is Saul and what is Anna, and between them and the other people in the book." Here is a merging more complete than that of lovers in Keats. Here is all the bliss of self-dissolution into an embracing whole that Lessing had sought—and failed to find—in Communism and in sex.

This romanticization of madness was very much part of the Zeitgeist in the early 1960s, and although Lessing—for once—rejects any suggestions of outside influence, she was close to the sources of the new psychology. Through Clancy Sigal she had entered the intellectual orbit of R. D. Laing, whose 1960 book, *The Divided Self,* attempted to redefine the borders between concepts of sanity and insanity, and at times seemed to award the advantage to the latter. (Sigal, who was a patient of Laing's and later wrote a book about the experience, liked to say that schizophrenics were the leftist intellectuals' newest substitute for the proletariat.) Lessing didn't need Laing to lead her to the idea that madness could serve as a necessary inner fortress, but Laing's notion that what we call mental illness may include a genuine experience of different and possibly higher dimensions of human awareness, dangerous only in that it exposed the "abnormal" individual to hatred and ridicule, seems to have had a profound effect on the fate of Martha Quest. In the final volume of the series, *The Four-Gated City* (1969), Lessing relinquished realism to have Martha join a secret network of mind readers and apocalyptic visionaries whose superior gifts make them appear to be mentally ill. When their direst predictions are fulfilled, the multibook saga that began with adolescent rebellion on a porch in Africa culminates in worldwide nuclear catastrophe, escaped only by those with the powers to foresee it. When last heard of, Martha Quest is rumored to be living on a remote island, refusing to reveal her whereabouts for fear of being rescued. Her position does not seem far off from that which Lessing had got herself into as a writer.

The large female audience that had been reading Lessing's novels to help sort out their lives—which is the way that Martha Quest herself read novels—would have gathered very little from *The Four-Gated City,* a huge, ambitious mess of a book, in which snippets of the old childish anguish ("mamma, why are you so cold, so unkind, why did you never love me?") extrude from mounds of lifeless narrative like roots that have been cut. The conflict between what Lessing wants to write and what Lessing needs to write finally results in deadly stalemate; at times it appears that apocalypse is just her way of evading personal problems. Yet on the "woman's issues" level, this book marks the first time Lessing presented her heroine as self-consciously middle-aged, thus opening a whole new area for grievance, and one that allowed her to bring the subject of mental breakdown back to familiar territory and to regain her audience in her next book, *The Summer Before the Dark.* Published in 1973, this widely popular novel made a case for understanding madness as a natural result of being a middle-aged woman and a mother.

A familiar indictment, perhaps, but it was issued now on vastly more sympathetic, first-person grounds, by an author who had reached her mid-fifties. Lessing's heroine is a London housewife whose children have grown and who discovers that she has been personally malformed not by the life she's missed out on—a common enough notion by that time—but, rather, by the way she has *spent* her years, in caring selflessly for others. "The virtues had turned to vices," Kate Brown berates herself, "to the nagging and bullying of other people. An unafraid young creature had been turned, through the long, grinding process of always, always being at other people's beck and call, always having to give out attention to detail, minuscule wants, demands, needs, events, crises, into an obsessed maniac. Obsessed by what was totally unimportant." This is also the view of the young woman Kate rents a room from when she runs away one summer, a young woman who is herself running away from a mother just like Kate, and from the ghastly fate she represents.

> "Her whole shitty life doing nothing, fuss fuss fuss about details, details. . . . I'm not going to be like you—it's my responsibility, saying no. I'm not going to be like my mother. You're maniacs. *You're mad.*"
>
> "Yes," said Kate. "I know it. And so you won't be. The best of luck to you. And what are you going to be instead?"

To her readers, this may be the most important question Lessing ever asked.

■ ■ ■

And what about answers? Perhaps that is too much to expect of a novelist, even one unafraid to make claims about the fate of the universe. Or perhaps, as it often seems, Lessing's writing thrives on the tension of irresolution; in that case, any clear answer would mean an unwelcome end to the search that itself keeps her going. After all her travails in *The Golden Notebook,* Anna Wulf pulls herself together, joins the Labour Party, and begins to do volunteer work to ease her need to make all things better than they ever can be; she also announces that she has given up writing. At a similar point of emotional recovery in Lessing's memoir *Walking in the Shade,* she announces a newfound allegiance to Sufism, and the reader almost sighs at the sound of the authorial boots tramping off uphill again, in this spiralling Pilgrim's Progress of a life.

But how far can a pilgrim go if every road is a circling back to unanswerable questions posed in a nursery a lifetime ago? It is not difficult to assess the importance of Lessing's childhood injuries in her continual search for union and belonging; it is harder to know how these unobliteratable memories have affected her achievement. Are her long-nursed hurts a secret strength, the source of her writing? Or have they been impediments to her getting her real work done? If this most fluid, productive, intellectually engaged of writers could be freed from the trammels of her family romance, what would she accomplish?

Lessing's most powerfully exorcising "attempt at an autobiography," as she called it, was published in 1974 and titled *Memoirs of a Survivor.* A lyric fantasy, it entwines the author's two essential subjects in the story of a woman who is mysteriously left with a child to care for as the civilization around her crashes down. The book may have been prompted in part by Lessing's one experiment with mescal, which seems to have stimulated visual memories like those revealed to the unnamed woman when she discovers that she can pass through the walls of her apartment in her search for a little girl whose crying voice she always hears. Beyond the walls is a glitteringly dry white nursery, in which she observes several cruel family scenes: a father torments his little daughter with surreptitiously sexual tickling and teasing, his actions revealed to the observing eyes to be—she had never been sure before—fully conscious and guilty; a tiny girl in a crib who has soiled herself and the surrounding whiteness is swept up in a whirlwind of scrubbing and near boiling by her furiously disgusted mother; the same girl's baby brother is taken in to say good night to their father while she—smelly, disgusting, female, badly behaved, why else?—is left behind, aching and forgotten. This is Lessing's "Jolly Corner," a ghostly search for the self she left behind or might have been. And it is an extraordinary achievement: despite the story's fantastic elements, the tone is natural, the people are true, the language is taut and glowing. An unknowing reader could be completely satisfied without recognizing the keys to so much of Lessing's earlier work. But a knowing reader is also happily surprised, for, despite its terrible record of childish pain, this is a work of warmth and forgiveness. When, at last, the woman catches a glimpse of the small, miserable figure for whom she has been searching through these rooms and scenes for what seems like all her life, it is not herself she beholds but her very young and weeping mother.

And this is Lessing's first book to boast a joyful ending, if not for the world then for her characters, a little parade of refugees who flee their ruined city together, with children tumbling after. Despite similarities to *The Four-Gated City,* the catastrophe here is social rather than nuclear, and the triumph of the ending is that the woman is finally able to leave her past behind—the lure of wandering through those ghostly rooms had become very strong—and to escape, just "as the last walls dissolved." The little group's destination is unspecified, but there is no doubt on anyone's part that they are following their leader ("that One who went ahead") into "another order of world altogether"—a phrase that occurs on the final page and may strike some of Lessing's readers as ominously familiar.

Where could they be going? In the books that followed, the new order of world turned out to be—to the old literary world's amazement—in outer space. Lessing published her first science fiction novel in 1979 and quickly added four more to complete the doorstopper *Canopus in Argos: Archives* series, which inspired more critical shock than should have been felt by anyone who'd read *The Four-Gated City* or, for that matter, by anyone who paused to think what a furiously willed sense of fantasy it required to remain a Stalinist through the mid-1950s. Lessing herself made this earthly connection clear. How big a leap was it, after all, from anxious Party discussions in Kensington kitchens to the invention of the interplanetary substance "SOWF," the "substance-of-we-feeling" that holds the virtuous Canopean empire together? How unfamiliar are the Canopean illnesses "Undulant Rhetoric" and "What Is the Point-ism"? And how unfathomably intergalactic is the heroine who abandons her children in fertile Zone Three, so that she may enter, unencumbered, the icy, ethereal heights of Zone Two?

You can take the woman out of the galaxy but you can't take the galaxy out of the woman. "Science is where our frontiers are" is how Lessing explained her flight into Canopean realms, and avowed that it was her ambition to write a novel that would take a reader to the new frontiers of consciousness—as *Moby-Dick* and *Wuthering Heights* and *The Story of an African Farm* had done for readers in the nineteenth century. It was this desire to write a book to which people could look for "news about humanity" (really no different from the impetus that had set her to *The Golden Notebook*) that seems to have carried Lessing through some twelve hundred pages of overstrained space-age social criticism. She was nothing if not earnest, even dogged, in carrying out her intentions, but there is far more "news about humanity" in the two volumes of her collected stories published during this period—*African Stories* and the even more exceptional non-African *Stories*—both of which show the author at her down-to-earth best.

Despite her theories and her ethics and the range of her literary personae—the African realist, the London scene painter, the anguished psychologist, the social prophet—Lessing is in essence a storyteller, with a rare gift for getting characters on their feet and for setting the wind stirring the curtains with language so apparently simple it betrays no method at all. The classical concision of the story form seems to induce in her an unusually clear-eyed mental energy, an urge to pick the locks of the elaborate cages she constructs in her novels. The exquisite "Homage for Isaac Babel," for example, renders adolescent tenderness and a psychology of literary style in three perfect pages; "The Day Stalin Died" and "How I Finally Lost My Heart" divide the weighty themes of *The Golden Notebook* and release them like birds into the air. These stories seem to have written themselves. The persistently autobiographical, tormentedly self-conscious novelist seems finally to have succeeded in losing herself—the lightness of self-forgetting buoys the results—in her own work.

Lessing literally disappeared as an author when, in the early eighties, she published two novels under the pseudonym Jane Somers. She was not slumming in low-down genres; quite the contrary. Aside from trying to prove something about the injustices of the literary establishment—two of her longtime British publishers rejected the manuscripts, a fact she gleefully announced once the hoax was exposed—she was trying to get free of the restrictions she now felt being "Doris Lessing" imposed on her writing, to escape what she termed "a kind of dryness, like a conscience," that marked everything she did. And there is a freedom from what might be called public conscience in the twin novels that were eventually published as *The Diaries of Jane Somers,* in that they are almost scornfully apolitical: "They don't know it's their social lives," one character remarks of a group heading off to a meeting, "they really believe it's politics," an observation that echoes down the coffee-cup-cluttered meeting halls of so many other Lessing books. And yet, these pseudonymous "little books about the emotions" may be Lessing's closest explorations of her quintessential subject: how to break down the shell of isolated selfishness, how to stop looking in the mirror, how to love. And now, explicitly, how to love not like a lover but more like a mother—that is, to care even for those who are difficult or unbeautiful and whose reciprocal emotion confers no worldly advantage.

Lessing ventures closer to an answer on these questions than she does on the future of women, but it turns out to be the very answer she was trying to escape. For the way to blessed self-surrender is, as ever, through the conscience: albeit, now, through the private conscience, which is perhaps more difficult to betray than the political variety. Jane Somers, latest heroine and alter ego, is a childless woman approaching fifty, a coolly efficient London fashion editor—Lessing described her as "what my mother would be like if she lived now"—who has watched both her mother and her husband die of cancer and felt nothing. The immaculate Jane shocks herself as much as everyone around her when she reaches out in several acts of effortful, dirtying, and sustained devotion: most overwhelmingly, to an angry, lonely, incontinent, dependent ninety-two-year-old woman, who spends an entire novel refusing to die. ("Of course," Jane's niece informs her, cutting the reader—and the biographer—off at the psychological pass, "she's just a substitute for Granny, you weren't nice to her, so you are making it up.")

Lessing fills in the allegorical shape of this story with compelling, nearly hyperrealistic detail that evokes the extremes of traditional female experience: from a silky pleasure in dauntingly useless ornamental beauty to the physical care—brute hands-on care, with hot water and a brush and a hidden gulping down of repugnance—for what beauty leaves behind. With regard to the latter, this book seems to effect an undoing of the awful nursery-shame recorded in *Memoirs of a Survivor.* Jane's adopted elder friend is called Maude, like Lessing's mother, and the care she requires is not so different from that required by a baby. But the mother love that Jane teaches herself to display and finds she really feels is patient and unrepulsable and irrevocable, a balm for old pains and a recompense for care withheld, from mother to daughter and back again in an eternal cycle.

The emotional testing of the ultimate bond was brought to the breaking point in Lessing's masterly 1985 novel, *The Good Terrorist*—even terrorists need mothers, perhaps more than most—and then carried far beyond in a slender fable of 1988, *The Fifth Child.* A modern maternal nightmare akin to *Rosemary's Baby,* it tells of an inhumanly begotten monster-child who nearly tears apart his pregnant mother's body before fully destroying the pleasant suburban family into which he is born. Despite the story's otherworldly suggestions, one can easily locate its nonsupernatural origins in the adversarial hell of Martha Quest's six-page labor, and in the raging battle for survival that exists between all of Lessing's mothers and children. Here the winner is clear: the mother in *The Fifth Child* sacrifices everything, including her marriage and the love of her other children, to save and keep a creature presented as little more than a force of evil in a childlike body. Yet for her there is no other choice. She is needed; no one else will bear the responsibility; she cannot turn away. It is clear that Lessing has come full circle, to a vision of maternity as obdurately binding as it had once seemed unnatural and imposed. Certainly, the cost for the mother is no less—the cost could not possibly be more—but there is no longer any doubt that it must be paid.

In her memoirs, Lessing defends having sent her son Peter to boarding school when he was twelve, and mother and son have evidently remained close. *Memoirs of a Survivor* was dedicated to "my son Peter," and Lessing depicts him strolling into the room to suggest that she write an adventure story about an orphaned brother and sister—it happened to be the very thing she was already writing—at the start of her newest novel, *Mara and Dann* (1998). Lessing mentions her older son in her memoirs just twice: when, coming through London for a visit at the age of eighteen, he discovered that she was not "Hecate incarnate," as he had been taught; and later, as a middle-aged man, telling her that he understood now why she'd had to leave his father, but that he still resented it. There is no mention at all of her grown-up daughter, who is last observed in the memoirs as hardly more than a baby, in Africa, around the time Lessing left her with a neighboring lady for a month so that she could make a trip down to the Cape. The neighbor had no children of her own, and Lessing returned to find her daughter delightedly spoiled with kisses and attention. She reports that the woman could not stop thanking her, and wept when Lessing took her daughter home, and told the young mother she thought she must be mad to have given up even an hour with such a perfect baby girl. All Lessing adds is that now she thinks so, too.

Hearts and Minds

Hannah Arendt and Mary McCarthy

Declaring pity for Hitler, in the spring of 1945, proved to be more cleverness than even the company around Mary McCarthy could bear, and the Manhattan cocktail party gave way to a scene. Hannah Arendt, who had got out of Germany in the same year that McCarthy got out of Vassar, demanded to know how McCarthy dared say such a thing "in front of me—a victim of Hitler, a person who has been in a concentration camp!" But of course McCarthy had not intended a serious moral or political statement. In fact, she held no strong opinions about the war, as yet, beyond a vague sense of accord with other *Partisan Review* thinkers that it was none of "our" business. Her intellectual celebrity and dauntless cultural criticism notwithstanding, McCarthy's politics were cliquish and literary, and her attention to the devastation of Europe had been confined to appraisals of war movies and war correspondents. Her point that evening was a matter of psychology: it was Hitler's longing for the love of the Parisians, even while he occupied their city, that struck her as so uncomprehending as to be pitiable. McCarthy's usual audience would have been able to hear in this display of ingenuous wonder its author's well-known love of shock and contradiction, as well as the high-flown political imagery that marked her rueful autobiographical stories—stories in which wives collapsed from within like France and the unconscious was a Soviet prison and the message was always entirely self-conscious and personal. The message this time seemed to be that, unlike Hitler, Mary McCarthy knew better than to expect love from her victims.

It was not a message that Hannah Arendt was equipped to hear. Deaf to excuses, she stalked off to upbraid their host, Philip Rahv, for the kind of talk he allowed in his apartment. ("How can you . . . you, a Jew?") Thus began the standoff between two of the most intellectually and temperamentally opposed women anybody knew, which went on for almost four unspeaking years, until Arendt approached McCarthy on a cold subway platform after a meeting for a political magazine and uttered the propitiatory words "We two think so much alike." From this richly productive lie grew one of the literary world's most loyal friendships, lasting until Arendt's death a quarter of a century later. Shortly before McCarthy's own death in 1989, she proudly helped to collect and edit the letters that form the record of this improbable alliance. And it no longer seems, after all, so improbable. Anchoring the bouts of gossip and philosophy, the professional outrages and the marital arbitrations of *Between Friends: The Correspondence of Hannah Arendt and Mary McCarthy* is the conviction of two erect and stubborn souls that the life of the mind and the communion of friendship have been their only means of survival. Emerging from their distant corners of experience, these women shared not a way of thinking but a deeply aggrieved and implacable sense of justice, and a perfect willingness to play the judge.

"I've read your book, absorbed, for the past two weeks, in the bathtub, riding in the car, waiting in line in the grocery store," McCarthy declares at the start of her first letter, in 1951. The golden girl of the smartest set, the literary mistress of trysts in upper berths and morning-after abashment, McCarthy had recently fled New York, married for the third time, and was claiming to have more serious matters on her mind. The volume balanced on her soapy knees and accompanying the calm domesticity of her days in a rambling house in Rhode Island was *The Origins of Totalitarianism,* 477 pages hardbound, a book that Arendt had begun writing in 1945, after a decade of assembling materials, and after finally overcoming her sense of "speechless outrage and impotent horror."

Arendt's original, driving aim, openly declared in the early outlines of the book, was to argue against the inevitability of history, and to show that the Nazi catastrophe could have been avoided. The unsettling aspect of this argument was her insistence that the Jews themselves must face up to a significant measure of responsibility for their fate. In the completed work, Arendt charged the Jews with chronic political indifference and, historically, with having created the conditions of isolation and self-mythologizing racial antagonism that were to be turned against them to such devastating effect. To her mind, there was a fatal continuity between the notion of a chosen people and that of a master race.

For the reading public, Arendt's point about Jewish history was diffused and her fury generally disguised by her book's imposing armature of scholarship and its dazzling explorations of anti-Semitism, racism, and nineteenth-century imperialism, the forces that in her view had prepared for the twin totalitarian states of Hitler and Stalin. In fact, it was the very pairing of these states, in 1951, that most seized attention: the immediate impact of the book was as a kind of glorified Cold War tract, providing intellectual ammunition for the view that Soviet expansion could be stopped only by the methods that had finally stopped Hitler. This was not a conclusion Arendt had intended, nor did the particulars of her research or analysis bear it out. Compared with Nazi Germany, the Soviet Union is treated in a notably cursory manner throughout *The Origins of Totalitarianism.* The fact that anti-Semitic ideology was in no way as important to Stalinism as it was to Nazism did not deflect her from her focus on "the Jewish question" or her theory of the systems' common origins. About economics or Marxism as a historical force she had little to say; she had planned to write a chapter on these subjects but never completed it. For better and worse, the history Arendt produced was the history she felt compelled to think about.

Perhaps the most personal stroke in Arendt's vastly learned study, however, was her refusal to consider German cultural history among the "origins" of modern political catastrophe. Brushing aside the frequently cited connections—in 1945, for example, Thomas Mann had advised the German people to examine their intellectual patrimony, from Herder to Heidegger, in order to confront the delusions responsible for National Socialism—Arendt exonerated her beloved *Kultur* from a primary role in the development of Nazi anti-Semitism. Instead, she located the Nazis' ideological sources in the institutionalization of racism in colonial Africa, a precedent for which, as she pointed out, all major European states shared responsibility. Further, she claimed that exclusionary race theories would have resulted from the African experience without any ideological preparation at all, since such theories were a natural reaction on the part of shocked and terrified Europeans—in particular the Boers—when confronted with what seemed "the overwhelming monstrosity of Africa" and its vast population "as incomprehensible as the inmates of a madhouse." Reading Arendt, one almost begins to feel sorry for the Boers. More, one feels caught up in an intense authorial compulsion—passionate, intelligent, persuasive—to muddle and even reverse the historical roles of victim and oppressor.

Clearly, Arendt did not write a balanced book, or an objective one. When McCarthy went on to say that she found the book "engrossing and fascinating in the way that a novel is," it was not mere hyperbole or a convention of speech but a typically frank insight, and unusual for its time. Surely one of the most stimulating aspects of Arendt's book—even for readers who do not agree with its premises—is its presentation of large ideas and quantities of information in a style that is richly textured and metaphorical, a style more associable with belles lettres than with standard political analysis. In fact, Arendt's immensely powerful and tragic book is as much a work of literature as a study of history, with its infernal echoes of Conrad and Kafka, Orwell and Yeats rebounding upon a hell that men had actually built.

Late in the book, Arendt pulled her themes together in a burst of metaphysics, concluding that beneath their differences the totalitarian states were profoundly united by an unprecedented form of "radical evil." Although she had borrowed the term from Kant, Arendt insisted that it now served to describe an entirely new category of evil—inexplicable by even the basest human motives of greed or desire for power, existing in and for itself alone. ("The object of persecution is persecution," Orwell had instructed in *Nineteen Eighty-four.* "The object of torture is torture. The object of power is power. Now do you begin to understand me?") Such evil was manifest in the anonymous terrors of the modern police state and, above all, in that state's most representative achievement: the concentration camp. There, evil's most radical victory was won, Arendt argued, not in the torture or murder of innocent victims, which has always been a part of history, but in the programmatic destruction of the victims' individuality, their sense of conscience, and finally their innocence itself. For Arendt, the Nazi co-opting of camp inmates to oversee and punish other inmates, serving as "Kapos" within a chain of command, meant that the Nazis had succeeded in making their victims complicitous on the highest moral level. In the camps, she wrote, "the distinguishing line between persecutor and persecuted, between the murderer and his victim, is constantly blurred."

For readers of the concentration camp studies of Bruno Bettelheim or the memoirs of the French Resistance fighter David Rousset—Rousset is an obscure figure today, but his books were Arendt's most frequently cited source of "inside" information—there was nothing new in these facts of how the camps functioned or in the corruption of the Kapos, or, for that matter, of the corruption of many other inmates who struggled to survive. ("Victim and executioner are alike ignoble," Rousset wrote; "the lesson of the camps is the brotherhood of abjection.") Nevertheless, Arendt's use of insights and images from Rousset's books is at times marked with shifts of emphasis that create a meaning different from that of the original. For example, Rousset's observation that in the camps "nothing is more terrible than these processions of human beings going like dummies to their death"—terrible, he specifies, because of the intolerable awareness of the oppressors' power that is forced upon those who watch—is adapted by Arendt into a furious indictment of the victims. "Nothing remains but the ghastly marionettes with human faces," she writes, "which all behave like the dog in Pavlov's experiments, which all react with perfect reliability even when going to their own death, and which do nothing but react." The change is linguistically subtle but emotionally enormous: it is no longer the sight of humans transformed into dummies that is terrible or ghastly, but the "marionettes" themselves. In a larger and less subtle instance, Arendt chose to ignore Rousset's extensive tribute to inmates who had held fast to personal codes of honor, those heroes and martyrs of his personal acquaintance—among them even Kapos—to whom he dedicated the very book from which Arendt took the Nietzschean epigraph of the final section of *The Origins of Totalitarianism:* "Normal men do not know that everything is possible."

In the way she selected and shaped her information, then, and in the essential lines of her thinking, Arendt emphasized the absoluteness of the Nazi victory in the grandest possible philosophical terms. The Nazis' obliteration of all sympathetic outside witness to their victims' plight, their manufacture of what Arendt memorably termed the "holes of oblivion" in which all moral acts were buried with the dead, made resistance meaningless and—for the first time in history, she claimed—martyrdom impossible. (This was another Orwellian concept: Philip Rahv, in an essay on Orwell, wrote in 1949, "What is so implacable about the despotisms of the twentieth century is that they have abolished martyrdom.") In conclusion, Arendt argued that by force of this "radical evil" death itself had been changed, from the completion of a life to its negation.

For Mary McCarthy, however awed by the sweep of Arendt's ideas, the very notion of such satanic omnipotence was unacceptable. In the midst of her praise for the book, she objects to the way Arendt seems to credit the Nazis with a deep political knowledge and an ability to interpret their times "as a great maître would have done," in McCarthy's slightly self-consciously Europeanizing words. With bright-eyed practicality, she insists that many features of these regimes fell into place "without anyone's cleverness, simply because they worked." Instinctively, it seems, McCarthy hints at a revolutionary notion that would explode a decade later, with *Eichmann in Jerusalem* and Arendt's formulation of "the banality of evil"—a notion that would overturn many of the moral conclusions of the earlier book. Here is a subject that continually resurfaces in these women's letters, in one form or another, as it does in so much of their work: the sources and powers of evil, the possibility of goodness, the durability of the moral impulse. It does not subtract from their intellectual or imaginative achievements to suggest that for both women these matters were personal, obsessive, and suffocatingly close.

Arendt's own history has made her credentials if not her compulsions as a moralist very clear. Born into an assimilated Jewish family in Hannover in 1906, brought up on Goethe, she existed in an unstable, oxymoronic condition—German Jew—that she would come to believe only Heine had mastered without fraud or bitterness. Her Russian forebears had immigrated to Königsberg, capital of East Prussia, as a center of Enlightenment opportunity; indeed, they thrived in business and saw their children educated at the university where Kant had taught. Her father acquired a degree in engineering and a passion for Greek and Latin literature; her mother was sent to study music for a time in Paris. Soon after their marriage, Paul and Martha Arendt decided to risk having a child, despite their knowledge that he might pass on a syphilitic infection he had contracted years before. Hannah, their only child, was born healthy, but her father's illness overtook him again when she was two and a half.

From then until his death, when Hannah turned seven, the little girl saw and experienced what her mother described as Paul Arendt's "entire horrible transformation," helping to care for him at home and finally in the hospital, up to the time when he could no longer recognize her. In Martha Arendt's journals, cited by Elisabeth Young-Bruehl in *Hannah Arendt: For Love of the World,* she noted that little Hannah was a model of patient tenderness with her father, only occasionally betraying a wish that he were already dead. The child prayed for him often, although she had not been taught to do so. Hannah Arendt's own later characterization of her youth as "fatherless" seems unjust only in being too static to convey her childhood's prolonged, consuming act of parting. One can only wonder when and how she learned the nature of her father's illness, and whether this affected her evident tendency to believe that people bring their own worst fates upon themselves.

Arendt always stressed the lack of religion in her background. Her parents had returned to Königsberg when her father became ill, and as a young child she was sent to synagogue with her grandparents, and had even once announced her intention of marrying the rabbi who gave her elementary instruction—which was provided by her family as a supplement to the city's mandatory Christian Sunday school. Yet she claimed to have discovered her Jewish identity only through the insults of the outside world. Königsberg had become a major stopping point on the railway line running up from Odessa, and the ever-renewed influx of poor and peculiar-seeming foreign Jews was a source of general social friction and of pained embarrassment to those who had edged into assimilation. Hannah Arendt's mother taught her to walk straight and silently out the schoolroom doors when subjected to a teacher's anti-Semitic taunts; when taunted by other children, however, she was to stay and fight. Decades later, she recalled that it was by keeping to these rules that she had learned to preserve her dignity, and had managed to prevent her soul from being poisoned while learning what it meant in this world to be a Jew.

Neither history nor politics nor anything to do with "the Jewish question"—a subject she found "boring"—were among her early interests. At fifteen she organized a Greek reading circle among her schoolmates, and she studied Christian theology when she went off to the University of Berlin. It was probably there that she first heard of Martin Heidegger, who had been creating a stir since his appointment at the University of Marburg in 1922. "There exists a teacher" is the response she recollected; "One can perhaps learn to think." By the fall of 1924, when Arendt enrolled at Marburg, Heidegger had become the center of a cult of students who saw him as the new apostle of the German philosophic tradition—itself a compound of romanticism and theology in which the learned apostle gladly stood in for the God that German philosophy had killed. Heidegger's students were a close-knit group, reportedly gathering after every lecture to try to figure out exactly what the master had said.

Although Arendt excelled at comprehending the great teacher's lectures, she was perhaps less astute at interpreting the personal communications that were soon coming her way. Heidegger's first letter to "Frau Arendt" was sent in February 1925 and praised her mind and soul while bemoaning the loneliness of the scholar's life; a note to "Dear Hannah" written two weeks later makes it apparent an affair had begun. She was eighteen years old, one of only two Jewish students in his classes. He was not only a celebrated genius but a self-styled Romantic figure out of Caspar David Friedrich, a German magus who dressed in Black Forest peasant clothing, a small and dark man whose face was said to be difficult to describe since he could not look anyone in the eye for long. He was thirty-five and married, with two small children. Several of the letters that Arendt eventually left to the Deutsches Literarchiv in Marburg—letters that produced a sensation when Elzbieta Ettinger published *Hannah Arendt, Martin Heidegger* in 1995—are devoted to details of their clandestine meetings, to places and times and to an elaborate signal system of lights switched on and off, none of which served to dissuade the young student from the idea that she was involved with anything but "philosophy incarnate."

"Volo ut sis"—"I want you to be"—he wrote her early on, quoting St. Augustine in testament to his feelings for her. Arendt would quote the phrase herself a quarter century later, in *The Origins of Totalitarianism,* to suggest the only nonpolitical way by which the effects of natural human inequality may be overcome: that is, "by the great and incalculable grace of love, which says with Augustine, '*Volo ut sis* (I want you to be),' without being able to give any particular reason for such supreme and unsurpassable affirmation." Arendt's own understandable sense of inequality is clear in the affirmation she offered Heidegger during the three years of their affair. Beyond the clichés of the eroticized master-student relationship, or the Pygmalion-Galatea relationship, or even the German-Jewish relationship, he presented her with a paradoxical dream of living fully and "concretely" in a realm of pure thought. "Heidegger never thinks 'about' something; he thinks something," she reported decades later, still poised between metaphysics and infatuation. Among the philosophically minded, this is not an uncommon sort of praise, usually meant to emphasize the real-world grasp and significance of a highly abstract thinker. ("He pre-

sented things, not generalizations about things," Alan Bloom wrote of Leo Strauss.) But Arendt's statement was also a reflection of Heidegger's philosophic project, his attempt to ground philosophy in the concrete experience of "being-in-the-world." The lectures Heidegger was giving in these years resulted in *Being and Time,* perhaps the most influential and least concretely fathomable of modern philosophic tomes. The book established his international reputation when it was published in 1927, setting off existentialist sparks all over Europe. At Marburg, it earned him a full professorship, a position that may have increased his awareness of the risks he was taking in his private life.

Arendt had actually left Marburg for the University of Heidelberg in the spring of 1926, to begin her dissertation with Heidegger's avuncular colleague, Karl Jaspers. There were practical reasons for the move, which was made at Heidegger's behest, as both her work and his reputation stood to be compromised were their relationship discovered. Even so, the terms in which he advises her to go are harsh. When he accuses her of failing to fit in at Marburg, despite her academic success, one cannot help sensing the fears he must have been exploiting. Once ensconced in Heidelberg, Arendt continued to meet Heidegger whenever he sent word—she was apparently not allowed to write him during this period—although his requests eventually fell off until their meetings took place only several months apart. Heidegger finally broke off the affair entirely, in early 1928, at a time when he was penning *"Volo ut sis"* in a letter to one of his wife's more attractive friends.

Arendt's anguish during these years is apparent not only in her letters following the break but in the dissertation that she was writing, on "Saint Augustine's Concept of Love." If the subject of Augustine was owed to Heidegger, the most powerfully human of saints was nonetheless a figure of direct and inevitable appeal for Arendt, who wrote of him as a product of dual historical citizenship, Roman and Christian, whose life "in between" allowed him to comprehend opposing worlds. When her dissertation was published, in 1929, reviewers found it highly original but perhaps misleading in its emphases on certain unrepresentative aspects of Augustine's thought. In an English version that Arendt expanded and then abandoned in the 1960s, which finally appeared as *Love and Saint Augustine* in 1996, her Augustine sounds startlingly like his great confessional descendant and cotheologian of desire, Marcel Proust: "What makes love, defined as desire, unbearable," Arendt paraphrases, "is the constant fear—that must accompany love—of losing its object." It was this constant fear attached to human desire that spurred the pagan sensualist's conversion to the love of God that alone could not be threatened. All forms of merely human desire "must end in self-oblivion," Arendt explains, in "losing the holding-back power, the reference back to the self and its present notion of happiness." It is Augustine's important lesson that "all desire craves its fulfillment, that is, its own end. An everlasting desire could only be either a contradiction in terms or a description of Hell."

In September 1929 Arendt sought to cut off her own desire by marrying Gunther Stern, a former student of Heidegger's, whom she had known at Marburg. The pair had renewed their acquaintance just that winter at a masquerade ball in Berlin, held at—of all places—the Museum of Ethnology; Arendt had turned out as an ethnologically minded Arab harem girl. The costume alone suggests how hard she was working at putting the past behind her. And yet, shortly after her marriage she wrote to Heidegger to say that it was still he who provided "the continuity of my life, the continuity of our—*please* let me say this—love." Heidegger had come to pay the newlyweds a visit; Stern was apparently not aware that his wife had been anything but another admiring student. When the men departed the city together, Arendt secretly followed them to the train station. "I was standing in front of you for a few seconds . . . you did not recognize me," she soon wrote to Heidegger, and went on to relate a harrowing childhood memory in which her mother had read her a story of a dwarf whose nose grew so big that no one recognized him, and then pretended that this had happened to Hannah. "I can still remember the blind fear when I kept crying: 'But I'm really your child, I'm really Hannah.' Today I felt the same way." When the train rushed away, it was "exactly as I have imagined: the two of you high up, above me, and I alone, utterly helpless. As always there was nothing I could do but let it happen, and wait, wait, wait."

It was just about the time of this visit, in October 1929, that Heidegger wrote a letter to the Ministry of Education, entirely unprompted, warning of the "growing Judaization" of German spiritual life. Arendt had no way of knowing this, of course, although her summoning of the memory of the dwarf with the disfiguring nose does seem a roundabout and chilling way of asking—given the proliferating stereotypes in the German press—whether it was her Jewishness that had cost her his love. Doubtless she did not think that she was asking this, and the mature Arendt would be incensed at such an interpretation—all her life she had nothing but scorn for what she summed up as "Freud." But it is impossible not to see that in the memories and images she presented to him she had become, in her own mind, Heidegger's child, and that the confusing, inexorable loss of his love seemed a fulfillment of the fate begun with the death of her father. At the time, the two losses seem to have become almost indistinguishable, except that now she had no one with whom to share her bereavement, and that she could no longer pray.

Arendt certainly did know of the extreme Nazi sentiments of Heidegger's wife. Elfride Heidegger, an early fan of *Mein Kampf,* had invited Gunther Stern to join the Party's local youth group in 1925, before he'd informed her that he was a Jew. It was perhaps through the influence of Stern that Arendt's scholarly interests were changing: in 1929 she decided to write her second required thesis on the Jewish salons that had been home to German intellectual life in the late eighteenth and early nineteenth centuries. Arendt's resulting study of Rahel Varnhagen—a leading Berlin *saloniste,* originator of the "Goethe cult," and a friend of Heine's—is written in the form of an autobiographical meditation, something of a German-Jewish, feminine *Confessions.* Relying heavily on Varnhagen's own extensive letters and journals, Arendt begins with her subject's actual deathbed testimony, a spiritual reconversion made at age sixty-two, in 1833: "What a history! . . . The thing which all my life seemed to me the greatest shame, which was the misery and misfortune of my life—having been born a Jewess—this I should on no account now wish to have missed."

Born in 1771, a member of the first generation of assimilating German Jews, Rahel Varnhagen—maiden name Lévin—possessed neither a history nor a tradition of her own, Arendt points out, since these had long been reduced to the insufferable squalor of the ghetto. Because she also lacked the viable commodities of beauty, money, or even a decent education, Rahel's only refuge from her self-declared "infamous birth," and her only relief from the daily dissembling of every detail of her life, was in the conversion of those details into purely abstract thinking. "By abstraction reason diverted attention from the concrete," Arendt explains in a passage that makes her affinity for Rahel seem as natural as her attraction to Augustine, or perhaps even to Heidegger, "it transformed the yearning to be happy into a 'passion for truth'; it taught 'pleasures' which had no connection with the personal self."

Like Arendt's Augustine, her Rahel Varnhagen is preoccupied with protection from unbearable losses. Like Augustine, she exists in a space between two clearly defined cultures—*"das Mädchen aus der Fremde,"* the girl from a foreign land or even another world, in a phrase from Schiller that Arendt applied to Rahel Varnhagen and, many years later, to herself.

But, unlike Augustine, Rahel could not successfully make the leap across; although she officially converted to Christianity when she married, in 1814, both true faith and the larger, Christian world remained closed to her. The "consuming, hopeless yearning" that Rahel suffered was only temporarily appeased by her hard-won "gift for abstraction"—and it was precisely this gift that led her to disaster. Turning her back on the everyday reality of politics, Rahel sought not to free her people from oppression but to free herself from her people. It was only when a wave of virulent post-Napoleonic anti-Semitism swept through Prussia that this exceptional woman was forced to concede that "her life also was subject to general political conditions." Arendt saw that such a woman in such a time could not be expected to have done better; but there could be no such excuses anymore. The clear moral of Arendt's story is that history must be understood—and acted upon, as politics—or it will take its revenge as a personal destiny. By the time Arendt was writing the later chapters of this study, in 1933, she had decided to cast her own political lot with the Zionists. The dream of assimilation begun with Rahel's generation was formally over.

In February 1933, after the burning of the Reichstag, Gunther Stern left Germany for Paris. That spring, Martin Heidegger became *Rektor* of the University of Freiburg, and welcomed the National Socialist government with a speech honoring "the glory and greatness of the New Awakening." Hearing reports that he was rejecting Jewish students, turning on his Jewish colleagues, and behaving in general like a loyal Nazi—all true or about to be true—Arendt wrote him an angrily questioning letter. She received a reply that denied every charge of anti-Semitism, complained of slander, and stressed the favor he had always shown to Jewish students, presumably including Arendt. The incomplete manuscript of "Rahel Varnhagen"—begun as her final requirement for a teaching appointment in the universities that now excluded her—was one of the few possessions Arendt carried out of Germany when she fled that fall.

"When one is attacked as a Jew, one must defend oneself as a Jew" was the new core of her belief about how she had to live. Joining Stern in Paris, Arendt went to work for Youth Aliyah, an agency that oversaw the transfer of Jewish children to safety in Palestine. She became part of a circle of intellectual exiles, mostly émigrés from Berlin, among them Walter Benjamin and a self-educated former Communist revolutionary named Heinrich Blücher. Seven years her senior, Blücher had a passion for causes—he was one of the few gentile Zionists she knew—and also a hoard of experiences earned on political barricades. He became Arendt's teacher in the sphere of practical politics and, emotionally, he began to provide an antidote to the devastating effects of Heidegger. Arendt's marriage had never been more than an escape from those effects and had long since evolved into a friendship; she and Stern were amicably divorced in 1936. "It still seems unbelievable to me that I can have both, the 'great love,' and retain my own identity," she wrote to Blücher, wonderingly, in September 1937. "And only now I have the former since I also have the latter. Finally I know what happiness actually is."

Arendt and Blücher were married in early 1940, four months before they were separately interned as "enemy aliens" in their country of refuge. If, in later life, Arendt was notoriously quick to condemn any sign of passivity on the part of people faced with persecution, she was also plagued by the memory of how she had guilelessly reported to French authorities as instructed, and how this obedience had led to her imprisonment among thousands of political undesirables in a camp at Gurs. She was held there for only a few weeks; ironically, it was Hitler's victory that won her freedom. In the chaos that followed the fall of France and the camp's change of command, she simply grabbed a toothbrush and walked out the front gate. Arendt herself estimated that this chance was seized by about two hundred women, out of seven thousand. Five years later, she would lie to Mary McCarthy about having been in a concentration camp; Gurs was not even a labor camp, and it was not run by the Nazis during her time. For those German Jews who stayed where they had been put, however, the undesirable part of their identity was soon switched from the national "German" to the racial "Jew," and the next way out did not come until 1942, on trains to Auschwitz.

Walking straight out the gate, as her mother had taught her, almost restored the pride Arendt had lost in having turned herself over in the first place. In the south of France she managed to find Blücher, and also her mother—Martha Arendt had been persuaded to leave Germany only after *Kristallnacht*—and she sailed to New York, in 1941, to start her life again. A stipend from the Zionist Organization of America enabled the immigrants to survive while Arendt learned English, and a letter she wrote to the editor of the German-language newspaper *Aufbau* earned her a job as a regular columnist. As a journalist, Arendt fought for two causes: during the war, for the establishment of a Jewish army raised from different countries to fight Hitler "as a European people"; after the war, for the establishment of Israel as a binational state, set in official federation with the Arab population. Seeing both causes fail, she did not willingly adopt any others. "I am not qualified for any direct political work," she concluded. Instead, in 1949, at forty-three, Arendt was taken up with the impossible task of understanding what had happened to the civilized world, and why, and of putting it all into a book. When the manuscript was finished, late that fall, she prepared to return to Germany.

Mary McCarthy was just then embracing a new and expanded life of ethical action—or, at least, of determinedly lesser solipsism. For years she had been far too smart to be taken in by crude American war propaganda, and too far to the left to succumb to any sentiment as bourgeois as patriotism. The war had become suddenly real to her only in the summer of 1945, when a British film documentary on the battle against Rommel moved her to tears. This was, of course, rather late, historically speaking, but she quickly organized a committee to send money and books to needy writers abroad. It is typical of McCarthy's ever-charming, ever-self-lacerating honesty that this account of her dumbfounding political obtuseness was put forth by McCarthy herself, the Mary Magdalene of moral consciousness.

McCarthy had always seen her life as a battleground of good and evil, a result of the infamously brutalized childhood she recorded in *Memories of a Catholic Girlhood.* She was six years old, at the end of October 1918, when the McCarthy clan—dashing ne'er-do-well father, pretty young mother, and four small children—boarded a train in Seattle, heading off to see the children's prosperous Minneapolis grandparents. Four days later, the train was met at its destination by a line of stretchers and wheelchairs; the influenza on board had been rampant. Roy and Tess McCarthy lived on a few more days; their children also came down with flu and pneumonia, but soon recovered. Mary, the oldest and the only girl, awoke from her fever in her grandparents' house. She was aware of being shown off to whispering visitors as an exhibit of pitiable fate, but she was told nothing of what had happened. It was several weeks before she surmised that her parents were not coming back, and that they must be dead. She later recalled that she was proud of having figured this out—she compared it to figuring out that there was no Santa Claus—and even somewhat relieved. It meant that she was clever, and that her cleverness was useful, since it prevented her from accepting false comfort or from expecting anything again.

The children were quickly deposited with an aunt and uncle out of the Brothers Grimm. For five years, living just a few blocks from her grandparents' house, she was fed a diet based largely on turnips and potatoes ("The family had a sort of moral affinity with the root vegetable"), forbidden to read for pleasure, and regularly beaten with a brush or a razor strop for such offenses as winning a prize for a school composition. It was announced that such accomplishments threatened to make her swell-headed, or to suggest that she might be better than those on whose charity she subsisted. Often a beating was administered simply as a preventative; and then, as the oldest, she was beaten along with any of the boys who'd done anything wrong, since she had failed to set the proper example. At times there were more inventive tortures, as when the children had their mouths taped shut at night.

Twice she tried to run away, but there was no place to go, and the children were too afraid of vengeance to risk telling their story. By the time their maternal Grandfather Preston travelled from Seattle for a visit, when Mary was eleven, the thought of escape had long been abandoned. His questions of simple concern—why wasn't she wearing her eyeglasses?—and then his indignation and finally his miraculous act of mercy in breaking up the household and whisking her away, were met with stunned bewilderment. "We thought it only natural that grandparents should know and do nothing, for did not God in the mansions of Heaven look down upon human suffering and allow it to take its course?"

No less than Arendt, McCarthy had felt the breath of evil on her cheek, and so helplessly close that it may have been she who better understood the numbed emotions of the survivor. Her physical escape took place just weeks after her grandfather's arrival; the rest of her required many years. ("Freud would have lived in vain," she wrote with typical breeziness, had she "not ended up sobbing on a psychoanalyst's blue couch.") It is almost too easy to see how the years of sanctioned brutality led to a nearly violent need to unmask moral dishonesty, and to an equally pressing if more surreptitious need to find something real and good beneath the mask. It was among her teachers that McCarthy sought and occasionally found the elusive real and good—like Arendt, she trusted that the mind would lead the heart—beginning at the Ladies of the Sacred Heart convent school, where she discovered her vocation.

"Like all truly intellectual women," McCarthy recalled of her first important teachers, "these were in spirit romantic desperadoes." The highly sophisticated, French-educated Sacred Ladies imparted a plan of salvation in which ordinary sinners were blithely consigned to the flames, but—and here they implicitly gave tribute to the infamously unchaste, sainted Augustine—more grandly Luciferan spirits like Marlowe and (especially) Byron wore halos of glittering possibility. McCarthy's vocational epiphany came when one of the strictest nuns accused her, aged eleven, of being just like Byron, "brilliant but unsound." That is how McCarthy records the accusation in *Memories of a Catholic Girlhood;* in a letter to Edmund Wilson written a few years earlier, the words scribbled in lightning are "brilliant but immoral." What is clear is that, from the start, writing was for McCarthy a kind of contest with the devil, a task of polishing to spectacular brightness all the things one shouldn't say.

From the sacred mysteries of the convent to the twittering operetta of Vassar to the New York cocktail circuit, McCarthy found herself increasingly burdened with heroic principles in a life far too narrow to live them out. As a writer, this discrepancy became her principal subject. Her heaviest weight of early conscience, much confessed in print, was a habit of falling into bed with men she didn't like and marrying men she didn't love, and all for reasons she couldn't fathom. At twenty-one she acquired her first husband, an actor named Harold Johnsrud, on an impulse she claimed to have regretted by her wedding night, although she admitted she did enjoy the monogrammed stationery and the table settings and the bridge parties the couple gave in their cozy East Side walk-up. Domestic trappings and rituals meant a lot, and she had some trouble giving them up when she left Johnsrud for another man and then left him to become a Greenwich Village Trotskyist and the belle of the intellectual left. ("It was not difficult, after all, to be the prettiest girl at a party for the sharecroppers.") She records that in this bohemian period she had sex with up to three men per day, a routine she presents as a feat of bookkeeping more than of anything else; her *Intellectual Memoirs* includes an entirely affectless disquisition on penis size.

In her spare hours she was making a name for herself with book reviews in *The Nation,* and then as the regular theatre critic for *Partisan Review,* a plummy job she owed—by her admission, and everyone else's—to an affair with one of the magazine's founding editors, Philip Rahv. McCarthy moved in with Rahv in 1937, and the cozy Trotskyist dinner parties they were soon giving—she was back on the East Side—continued almost up to the day she went out and married Edmund Wilson. Even Rahv, who was crushed, was also rather impressed.

Wilson was forty-two, portly and red-faced, accomplished and powerful, the unquestionable first-place ticket in the New York intellectual sweepstakes. The unblushing bride was twenty-five, and once again she claimed to know by her wedding night that she had made a dreadful mistake. The problem wasn't sex but the unavoidable realization that sex didn't mean what it was supposed to; the occasion only emphasized the emptiness. And this time there was the additional problem that the groom had got blind drunk and physically threatened her and accused her of having brothers who were Communist spies. Four months later, when she was pregnant, Wilson apparently did beat her up—it should be noted that there are those on Wilson's side who say she was invariably the instigator and managed to inflict her own damage—and then committed her to a psychiatric hospital. This was merely the opening siege in what the antagonists agree was seven years of extraordinarily productive warfare.

In her *Intellectual Memoirs* McCarthy expends a lot of effort trying to figure out what seemed quite obvious to everyone at the time: why she married Wilson. He'd "dragooned" her into it, she claims, or else she'd done it out of her infinite guilt at having "slept with this fat, puffing man for no reason, simply because I was drunk." She even gives some serious thought to the question of whether his prose style was better than Rahv's, as some in the *Partisan Review* circle snickered. And then she casually throws in the fact that her good Grandfather Preston had died suddenly, of a stroke, on New Year's Day, 1938. Five weeks later, she ran off with Wilson, whose settled, tweedy life in Connecticut seemed to offer what she calls a "coming home" feeling. But by some whopping miscalculation, she found she'd come to exactly the wrong home: not her grandfatherly shelter but the lair of the wolf she had once barely escaped, and now with no one to come to the rescue except herself. McCarthy pondered her self-imposed, potentially self-annihilating plight while she was living it—while she was married to Wilson—in the wonderfully astringent stories that Wilson himself urged her to write.

Forced her to write is how she liked to put it, however grateful she really was. It was Wilson who had insisted that McCarthy's literary gift was not merely critical but imaginative, and she tells how he shut her up in a room with a typewriter and refused to let her out until she'd produced a story. This account in itself rather resembles the classic McCarthy sex scene, minus the highballs: an act of ambiguous collusion falling somewhere between rape and salvation. "Cruel and Barbarous Treatment" is the first, aptly named story she wrote under lock and key, in the first month of their marriage, "straight off without blotting a line." An oversophisticated tale of adultery and boredom and social posing, it became the opening selection in *The Company She Keeps,* a sharp and often balefully funny book of loosely connected stories, published in 1942 and widely dismissed by McCarthy's former critical peers as "high-grade back-fence gossip."

It was the one about whisky and sex on an overnight train titled "The Man in the Brooks Brothers Shirt" that got everyone more than ordinarily riled, although the story is far more traditional than it seems at first: what begins as a tale of self-consciously smart, up-to-date female *libertinage* quickly unwinds into a tale of self-consciously smart, up-to-date female humiliation and self-loathing, all exhibited by a painfully analytic heroine who spares no one, least of all herself. ("This freedom of speech of hers was a kind of masquerade of sexuality, like the rubber breasts that homosexuals put on for drags, but, like the dummy breasts, its brazenness betrayed it: it was a poor copy and a hostile travesty all at once.") Delmore Schwartz's private title for this story was "Tidings from the Whore," which reflects both the literary jealousy and the overtly sexual resentment of many of the male intellectuals (even those she didn't use or sleep with) who'd watched McCarthy climb right past them.

Even her harshest critics, however, had to admit that the final story redeemed the author's penchant for mining the depths of social trivia. "Ghostly Father, I Confess" is a pocket autobiography of the McCarthyan soul, a scathing confession of sexual and emotional sin narrated by a St. Augustine in silk stockings. An interior monologue addressed by a theologically minded young woman to her psychiatrist, it concerns the horrors of living with a brutal, domineering husband and her attempt to discover why she doesn't run away. Here she contemplates the habits of mind that drew her to him in the first place:

> The romantic life had been too hard for her. In morals as in politics anarchy is not for the weak. The small state, racked by internal dissension, invites the foreign conqueror. Proscription, martial law, the billeting of the rude troops, anything at all, is sweeter than responsibility. The dictator is also the scapegoat; in assuming absolute authority, he assumes absolute guilt; and the oppressed masses, groaning under the yoke, know themselves to be innocent as lambs, while they pray hypocritically for deliverance.

Perhaps there was good reason, after all, for Hannah Arendt to believe that she and McCarthy thought so much alike. For both women, pity and especially self-pity were not to be borne; all claims to innocence, even—or, perhaps, especially—one's own, were suspect. When, in the same story, the events of McCarthy's horrific childhood are adduced as a reason for her behavior, she responds: "I reject the whole pathos of the changeling, the orphan, the stepchild." McCarthy might easily be speaking for Arendt, and even more so when she writes that she had inherited from her father (a dimly remembered, largely imaginary figure) "the twisted sense of honor that was always overpaying its debts, extorting from herself and from others the coin of unnecessary suffering to buy indulgence for a secret guilt, an unacknowledged shame." From a vantage point no higher than common domestic misery, McCarthy took on the same earthshaking questions that preoccupied Arendt:

> To know God and yet do evil, this was the very essence of the romantic life. . . . And, as they said, it could not go on. If you cannot stop doing evil, you must try to forget about God. If your eyes are bigger than your stomach, by all means put one of them out. Learn to measure your capacities, never undertake more than you can do, then no one will know that you are a failure, you will not even know it yourself. If you cannot love, stop attempting it, for in each attempt you will only reveal your poverty, and every bed you have ever slept in will commemorate a battle lost. The betrayer is always the debtor; at best, he can only work out in remorse his deficit of love, until remorse itself becomes love's humble, shamefaced proxy.

Even remorse gave out in January 1945, when McCarthy finally ran away from Wilson, taking their six-year-old son, Reuel, and checking into a Manhattan hotel. The infamously designing woman had signed over all her money to Wilson and had to borrow two hundred dollars just to hire a lawyer; the notorious femme fatale was no sooner free than she got herself married again, this time to a twenty-five-year-old fact-checker at *The New Yorker* named Bowden Broadwater—McCarthy was thirty-three—who was a follower of her career, who struck most people as decidedly effeminate, and who clearly represented a living antidote to everything she'd undergone with Wilson. But McCarthy's ability to break permanently with Wilson was based on a new and far stronger attraction, evident in her new disdain for Wilson's "literary"—that is, nonpolitical—approach to culture. More and more, she was looking to European thinkers for her emotional sustenance and her reading lists.

Arendt was not McCarthy's first continental crush, but she was her most valued and satisfying one, just what McCarthy's intellectual ardor required: a teacher, as pure and above suspicion as the Sacred Ladies who had first won McCarthy's faith, and like those holy sisters a formidable guide to principles for which one might truly and justly suffer. McCarthy was only six years Arendt's junior, but seemed far younger by virtue of being an American. (The childless Arendt tended to think of her American friends as children, and the behavior of some of those closest to her—McCarthy, Randall Jarrell, Robert Lowell—did little to discourage her impression.) McCarthy's early letters to Arendt suggest a nervous courtship. She fears that she is boring, she worries over her lack of erudition and gravity ("petty but at any rate human," she typically apologizes for an opinion), and she veers between her uniquely vivid common sense and a grievous attempt to put on something more *hochphilosophisch.* It is difficult to imagine McCarthy telling anyone else that the novel she was working on (*A Charmed Life*) concerned "the shattered science of epistemology." (A few years later, she would write in *Memories of a Catholic Girlhood* that her sadistic aunt's educational program "was in truth totalitarian: she was bent on destroying our privacy.") And Arendt plays her own role to the fullest; in one letter she blithely refutes what she calls "the oldest fallacy of Western philosophy," assuring McCarthy that "truth . . . is always the beginning of thought," and not the other way around, no matter what Socrates and all the rest—clearly another circle of her intimates—may have said. It does sometimes seem as though Mme de Staël were writing to Becky Sharp. But one of the deepest pleasures of this correspondence is to see how each woman comes to fill a role neither could have imagined: how in their years of crisis McCarthy becomes Arendt's most trusted intellectual defender, while Arendt becomes an advisor of genuine wisdom on affairs of the heart.

Randall Jarrell characterized the marriage of Hannah Arendt and Heinrich Blücher as a "Dual Monarchy," in which neither of the royal partners was more equal than the other. Alfred Kazin cherished the "open connubial excitement" of the couple's philosophical conversation, and paid homage to "the most passionate seminar I would ever witness between a man and a woman living together." Mary McCarthy was no less enthralled by the intellectual fireworks and mutual devotion of her Olympian friends, but for all her awe of Arendt, she saw the couple not as divine equals but as erotically balanced student and teacher. "In their intellectual relationship there was something fatherly, indulgent, on his side," she wrote, "and pupil-like, eager, approval-seeking, on hers; as she spoke, he would look on her fondly, nodding to himself, as though luck had sent him an unimaginably bright girl student and tremendous 'achiever,' which he himself, a philosopher in every sense, was content, with his pipes and cigars, not to be." These words are from McCarthy's eulogy for Arendt, a loving memoir that necessarily encompassed both husband and wife—since, as McCarthy put it, for her friend "Heinrich was like a pair of corrective lenses; she did not wholly trust her vision until it had been confirmed by his." And yet it was also McCarthy who whispered to her own biographer, Carol Brightman, the startling news—still startling when Brightman slipped it into the introduction to *Between Friends,* in 1995—that it was Martin Heidegger who had been "*the* great love affair" of Arendt's life.

Just one year later, Elżbieta Ettinger's sensational mini-book about the Arendt-Heidegger affair stirred a storm of responses with such tawdry or gloating tag lines as "The Nazi and the Schoolgirl" and "The Banality of Romantic Obsession." Yet the book's essential story was far from new. The facts of Arendt's youthful affair had been fully set out in Elisabeth Young-Bruehl's scrupulous and admiring biography, published in 1982. Young-Bruehl had also outlined Arendt's reunion with Heidegger, in Germany, in the early months of 1950, and had included Arendt's revelation that the great philosopher had told her she had been "the passion of his life"; Arendt's account of Elfride Heidegger's furious jealousy was also mentioned. Despite her exposure of these facts, however, Young-Bruehl discreetly concluded that Arendt's "loyalty to Blücher was not called into question." As proof, she cited a letter in which Arendt detailed the Heidegger drama to her husband, and told him she'd thought of him all the while it was going on.

Although Ettinger presented a far more complex (and heated) view of Arendt's emotional loyalties, the result of her investigations was actually the opposite of what the scandalmongers implied: there was no resumption of a sexual affair between the unrepentant Nazi and the former schoolgirl. The truth of what happened is rather harder to grasp, for the real point of Ettinger's book—a point that can now be confirmed by the published letters—is how deliberately and how well Hannah Arendt was used, politically.

After the war, Heidegger was stripped of his teaching position. His old champion Karl Jaspers—who had got through the war years very quietly with his Jewish wife—had urged Heidegger to recant his Nazi ties and, when Heidegger refused, had given testimony about the danger of allowing such a thinker to continue to influence the young. Removed from his duties, Heidegger suffered a brief nervous collapse and then retired to live in Freiburg, working in an isolated cabin in the Black Forest that his wife had built for him years before. Jaspers, writing to Arendt in 1949, angrily characterized Heidegger as having "an impure soul," one that "continues to live thoughtlessly in filth." To this Arendt replied with a bleak corrective: "What you call impurity I would call lack of character—but in the sense that he literally has none and certainly not a particularly bad one." Yet she went on to call Heidegger a bold-faced liar, a man who "fast talks himself out of everything unpleasant," and added—perhaps most importantly, as this marks her acknowledgment of the much disputed link between the man and his work—that his "whole intricate and childish dishonesty has quickly crept into his philosophizing."

Arendt's public assessment was even more devastating. She had stood firmly against the publication of Heidegger's books in America, and in a 1946 article introducing "existential philosophy" to readers of *Partisan Review,* she questioned "whether Heidegger's philosophy has not been taken unduly seriously simply because it concerns itself with very serious matters." It was his political behavior, rather, that she believed should be taken seriously. She reported that he had joined the Nazi Party "in a very sensational way" in 1933, and that, as *Rektor* at Freiburg, he had refused access to the university to his teacher and friend, Edmund Husserl, because Husserl was a Jew. (Actually, the widely rumored charge that Heidegger had officially banned Husserl was untrue; he had merely severed all their professional and personal contacts.) The importance of such acts was not in any mere personal failing, she argued, but in the fact that Heidegger's "entire mode of behavior has such exact parallels in German Romanticism." In his lack of responsibility, in his delusions of genius and his despair, Heidegger was, according to Arendt, "really (let us hope) the last Romantic"—an observation by which Arendt affirmed a relationship between German intellectual history and Nazism that, in the larger framework of *The Origins of Totalitarianism,* she would go to epic lengths to deny.

In late 1949, immediately on completing the manuscript of her book, Arendt went off to work in Europe for six months, on a mission to locate looted Hebraic books and artifacts for the Commission on European Jewish Cultural Reconstruction. She arrived in Freiburg in early February 1950 and sent a note to Heidegger at once—her first communication since 1933. In response, he came to her hotel to invite her to his home that evening, with apologies for his wife's absence; Arendt returned for a second visit the next morning, as he had requested that she meet Elfride, whom he had told about the past. The following day, Arendt wrote a letter detailing both meetings to Blücher, including the now famed account of Heidegger's confession of his passion, and his wife's near hysterical jealousy. To Heidegger himself, Arendt wrote that their meetings had been "a confirmation of a whole life," and had saved her from "committing the only truly unforgivable disloyalty." She returned for another visit in March, and by the time she left she seems to have determined on a course of action.

Back in the United States, Arendt began to oversee the translation and publication of Heidegger's books, even negotiating his contracts. She recanted her earlier position on the importance of Heidegger's ties to Nazism and—by logical extension—on the importance of moral responsibility as a component of philosophic thought. With a suspension of judgment she never accorded those who had failed to act nobly as *victims* of war and terror, she retracted her condemnation of Heidegger's ban against Husserl—which she still believed he had carried out—on the grounds that the act was not personally directed but was part of a general order; she also excused Heidegger's removal of the original dedication to Husserl in a 1941 edition of *Being and Time* with the astonishing words—written to the Reverend John M. Oesterreicher in August 1952—"I think he can prove that he did so only under strong pressure." Arendt became furious even with Jaspers when, in 1956, he pushed her to sever ties with Heidegger, although she had not seen Heidegger for years and his letters had ceased. But by then Heidegger had been fully rehabilitated, both as a professor and as the presiding genius of German philosophy, and he was no longer in need of her assistance.

Heidegger's motives couldn't be clearer. As for Arendt's motives, reams of dark speculation have failed to connect with the vital fact that her journey to Europe in late 1949—six months away, alone—was a response to a crisis in her marriage. Heinrich Blücher may have played the role of Arendt's teacher, but the facts are that he had not easily adapted to New York, had not learned English well, and had remained unemployed and apparently depressed throughout the postwar years. Arendt was often in the position of defending his inactivity to others—including her mother, who lived with them for a long and difficult period—and she supported the household with editing and other jobs while she wrote her book. In 1948 she took a two-month working holiday in New Hampshire to try to move her manuscript ahead; she came home to find that Blücher was having an affair with a young Russian woman in their circle. Young-Bruehl passes over this information lightly, with a reference to Weimar sexual sophistication. But it is difficult to escape the impression that for Arendt the experience was anything but light, and that the public nature of the affair, at least, involved some pointed cruelty on Blücher's part; Arendt's entire circle knew about it. The marriage continued with the young woman in their lives for several years, and with issues of jealousy and trust suddenly a looming part of the philosophical agenda.

These issues form a steady undercurrent in the letters Arendt wrote to Blücher during her sojourn in Europe, letters that betray both anxiety and loving manipulation. To a close woman friend, Hilde Fränkel, Arendt wrote about nervously deciding to send Heidegger a letter when she got to Freiburg; to Blücher she wrote only that Heidegger had appeared at her hotel as soon as she arrived. It does not make the rest of the drama less true—only, perhaps, more poignant—to emphasize that it was Blücher to whom Arendt wrote of how Heidegger had vowed that "I was once the passion of his life," and that it was Blücher to whom she wrote of Elfride Heidegger's jealousy, in a tone of detached if horrified amusement. Jealousy was an emotion that Arendt had decided, with Blücher's help, she was to be above; the task was apparently a bit easier now that she herself could play the role of the longed-for, younger woman. Yet by the time of her second visit to Freiburg, in March, the tone of detachment was gone, perhaps because Blücher had responded with nothing more than teasing and encouragement, or perhaps because of the purplishly seductive letters that Heidegger himself had begun to send. In any case, during the next few years Arendt began to characterize herself as the great philosopher's muse: the only person who could save him from the crippling influence of his wife and restore him to creativity. Reading her letters, one has the frightening sense that she had begun to believe this, as though in walking the edge of an old, extinct volcano she had toppled in.

Arendt saw Heidegger only once more, for a few days in the spring of 1952, and then not again for fifteen years. "The Martin legend," as Blücher called it, was fueled by what she needed to believe far more than by anything that had happened. One senses the yearning and the dislocation of her youth return like unlaid ghosts when, after her first visits in 1950, she replied to an insult of Elfride's about Jewish and German women by writing Heidegger her own Rahel Varnhagen–like self-assessment: "I have never felt like a German woman, and have long since ceased to feel like a Jewish woman. I feel like who I really am—'the girl from another world [*das Mädchen aus der Fremde*].' " The girl disclaiming her identity was then forty-three, and across the Atlantic was taken for a giant, even for something of a holy terror: "Her presence on the West Side was like Lear's on the Heath," Alfred Kazin vouched adoringly. But Arendt had her own dangerous appetite for giants. Kazin added a wondering testament to her intellectual purity and depth without fully understanding all that his statement implied: "She had devoted herself to Augustine because of a single sentence: 'Love means that I want you to be.' " If Heidegger was indeed the last of the German Romantics, it is only by virtue of outliving Arendt.

"I'm going to Europe, but I'd give it up like a shot if there were any hope of doing anything here that required my presence," McCarthy wrote to Arendt in 1953. Even more than Arendt—who had added a loyally conscientious American citizenship to her internal German one—McCarthy had become a woman casting about for a home. Wanting a sense of purpose, she'd even flirted with law school, but decided she could better serve the causes she believed in through her writing. What she actually wrote in these years, however, was a series of novels excoriating her friends for their failure to act on their own political beliefs. The resulting books—*The Oasis* (1949), *The Groves of Academe* (1952), *A Charmed Life* (1955)—are at once thin and diffuse, and suffer not so much from their confinement to a narrow world as from a sense that no world could really be as narrow as McCarthy made it seem. While each of these books contains some deadly accurate and funny barbs at the expense of most of her characters, all succumb to bland generalities or mawkishness when dealing with a figure meant to be made of better stuff. In McCarthy's view, moral integrity had become the exclusive property of one or two European intellectual saints—Arendt, Nicola Chiaramonte—or, worse, of an embarrassingly idealized version of the author who now frettingly modelled herself upon them.

"My novel is going ahead, but I have you horribly on my conscience every time sex appears," she confessed—she was always confessing something—to Arendt in 1954. "You are tugging at my elbow saying 'Stop' during a seduction scene I've just been writing." The scene in question is the liveliest part of *A Charmed Life*—"which is about bohemianized people and the dogmatization of ignorance," she assured Arendt, as well as about sex—and its drunkenly ambivalent participants are a barely disguised (as a psychiatrist) Edmund Wilson and, of course, McCarthy herself (as a blonde). The scene has the chaotic vitality of McCarthy's earlier bouts of sexual wrestling and recrimination, in which neither participant can explain the chronic, driving McCarthyan mystery: why the woman accedes to flesh she finds revolting and to an act that she is never sufficiently unself-conscious or uncritical to enjoy. Perhaps it was just the elusive release into unself-consciousness that McCarthy sought in her tireless

Punch-and-Judy ravishments. It is easy to draw from her novels the idea that good sex requires some of the same gifts as good imaginative writing, and that the critically withholding self can't perform either act very well.

McCarthy finally got to Europe, travelling with her husband for several months, in 1955; the marriage was already shaky, and when she suffered a miscarriage, in Paris, Arendt went over to tend her. ("You were so *extremely kind and good*," McCarthy wrote her afterward, "it broke the ice in my heart.") Arendt's own marriage had restabilized by the mid-fifties, when her husband came into his own with a university teaching job, and the most dangerous of outside liaisons seemed to be behind them. Always a woman of immense reserve, Arendt did not easily allow herself to be seen in trouble or in need, even by McCarthy. Visits and telephone calls must have made up for much that is missing on Arendt's side of their correspondence, for so extreme is the sense of her lofty distance from life's surface—from the casserole dishes and gardens and weather and (of course) sex that spill over in McCarthy's letters—that McCarthy's casual reference to an Arendtian opinion on double beds (apparently negative) comes as a bit of a shock. But the indulgence that Arendt could not claim for herself she could, apparently, supply to others in abundance.

During a few weeks in the autumn of 1956, Arendt progressed from shielded outsider to full conspirator in McCarthy's increasingly convoluted romantic affairs—she even rerouted mail to McCarthy's husband from a country where his wife was not supposed to be. When McCarthy's new flame, a London literary critic, turned out to be a pathological liar, a drunk, and a thief, Arendt offered wholly sympathetic consolation about loving such a damnable but attractive man, and offered comparisons with Brecht and Heidegger. Three years later, it was Arendt who gently brought McCarthy's loyal but unloved Husband No. 3 around to the idea of a divorce, after McCarthy had met an American diplomat named James West, a mature and responsible married man with three small children, who had suggested they tear their present lives apart and make a new one together.

Most of what has been cut from McCarthy's published letters to Arendt are her libelous carryings-on over West's then wife, who McCarthy believed should be dispatched to a secretarial job ("She's all right, he says, in an office") to ease the terms of his financial settlement. The stressful period of the paired divorces and remarriage shows McCarthy at her worst, transformed into a venal and strangely self-uncomprehending Mary McCarthy character. Arendt responds to all this with a demonstration of mighty grace, acknowledging McCarthy's anxieties but not the ugliness of her words, urging her toward compassion—for her ex-husband and, above all, for West's children—and seeming nothing else so much as a loving Jewish mother: "You know I worry and I also have somehow the firm conviction that as long as I keep worrying, things will straighten out." Nevertheless, after a year of frantic counsel and commiseration, Arendt had to miss McCarthy's fourth and final wedding, in Paris in April 1961, because the trial of Adolf Eichmann had just begun proceedings in Jerusalem.

Eichmann in Jerusalem is a reporter's account of the first major Nazi trial in Israel. It is also, for much of its length, a product of evident pain and anger. Like *The Origins of Totalitarianism*, Arendt's journalistic essay hovers somewhere between history and polemic, with swellings into poetry, and betrays the embattled pride of an author for whom being a German Jew was still a far from resolved condition. While in Israel for the trial, Arendt often felt miserable and alienated; letters to Jaspers deplore the "typical Galician Jew, very unsympathetic" who was the prosecutor, and the "oriental mob" outside the courtroom doors. (Jaspers, who fully accepted German Jews—like his wife and Arendt—as *echt* Germans, registered frank surprise at the existence of "almost incomprehensibly human Jews" among Eastern Europeans.) Still more ominously, before the trial was even under way, Arendt expressed her concern that the prosecution might fail to uncover some "very crucial aspects of this devilish business"—that is, above all, "the fact of Jewish collaboration."

In the very first paragraph of *Eichmann*, Arendt complains, almost comically, of the poor quality of the trial's German simultaneous translation, and of the "old prejudice against German Jews, once very pronounced in Israel." Within a few pages she is comparing the Israeli laws on marriage to the Nuremberg Laws; variations on the theme of "lambs to the slaughter" occur half a dozen times in her opening account of the issues at stake. But it was her contemptuous treatment of the Jewish councils' cooperation with Nazi authorities—a well-documented if complex and anguishing matter—that proved to be the truly incendiary element in Arendt's book, and that won it a still angrily debated renown.

The facts had already been set out in Raul Hilberg's monumental study, *The Destruction of the European Jews,* which served as Arendt's own primary source for this information. As Hilberg summed it up: in the ghettos of the East the Nazis had set up councils (*Judenräte*) composed of local Jews to serve as their intermediaries, and had charged them with keeping up workload quotas, with providing lists of Jewish families and apartments and, ultimately, with organizing deportations. The council members often belonged to preexisting Jewish civic groups, and assumed their new tasks to be necessary to the ghetto's survival: productive people were less likely to be deemed expendable, and every day survived was a day closer to the end of the war. But that end was long in coming, and in the Mephistophelean bargains that were struck, the councils often came to seem indistinguishable from their Nazi bosses. To some they seemed worse, being traitors to their own: the first act of the heroic fighters of the Warsaw ghetto uprising was to shoot all the members of the *Judenräte*. Hilberg was coolly clinical in setting out these facts, and implacable in his view of their importance: "For our understanding of how the Jews were ultimately destroyed," he explained, "it is essential to know the origins of the Jewish bureaucratic machine. The Jews had created that machine themselves." But Hilberg's densely detailed three-volume work also made a case for the widely varying character and achievements of different councils in different parts of the Reich. And he expressed an understanding of the impossible situation in which the members found themselves: "They could not serve the Jewish people without automatically enforcing the German will."

Eichmann in Jerusalem offered no such troubling complexity. Although Arendt was fully aware of what she herself, in a 1952 book review, had called "the terrible dilemma of the *Judenräte,* their despair as well as their confusion, their complicity and their sometimes pathetically ludicrous ambitions," there was no sign of such sympathy in her work a decade later. Mixing techniques of emphasis and omission—treating Hilberg in somewhat the way she had treated David Rousset—she stressed the culpability of these wretched councils while removing them from their equally wretched context of torturing choices and hopes. Arguing that without direct Jewish cooperation the number of Jews murdered during the war would have been drastically reduced, Arendt again mounted the accusation—this time, head-on and inescapable—that the Jews were largely responsible for their own destruction.

Arendt claimed to be astonished by the uproar that broke out over *Eichmann in Jerusalem,* which appeared in five installments in *The New Yorker,* in February and March 1963, and as a slightly expanded book that May. The Anti-Defamation League of the B'nai B'rith issued bulletins against it; the Council of Jews from Germany pleaded with her to stop publication. Innumerable articles claimed that by exaggerating the complicity of the victims and emphasizing the ordinariness of the defendant—the book was subtitled *A Report on the Banality of Evil*—she had written a defense of Eichmann.

By the fall, reactions had reached a level that McCarthy, with her abidingly delicate sense of political metaphor, termed "a pogrom." Urging Arendt to defend her ideas, McCarthy received the testy reply that "there are no 'ideas' in this Report, there are only facts with a few conclusions." But along with her reply Arendt enclosed a closely typed four-page memorandum in which she refuted the charges brought against her and suggested that McCarthy might do something about the nasty business herself.

She could have found no more congenial candidate for the job. The concept of "the banality of evil" which Arendt had derived from the cliché-ridden person of Eichmann was prefigured in McCarthy's startlingly similar account of the character of Macbeth—"the only Shakespeare hero who conforms to a bourgeois type: a murderous Babbitt"—in a compelling 1962 essay that Arendt had read and warmly praised while working on "the portrait part" of her Eichmann book. But then, McCarthy might well have derived the idea for her prosaic Macbeth from the coverage of the Eichmann trial itself. For Arendt was far from alone in her reaction to the banal appearance and demeanor of this giant of evildoing. Reporters at the trial seemed to vie with each other for the homeliest comparison: slight and pale and twitchy and sniffling, S.S. *Obersturmbannführer* Eichmann seemed, to their amazement, just like a milkman or a dentist or a small-town tax collector or the Vacuum Oil Company travelling salesman he had actually been before the war—a perfectly ordinary-looking monster. Unlike so many others, McCarthy found nothing in this to put her philosophy out of joint. She had known the type since she was six years old.

Add to this the fact that in August 1963, after eleven years of work, McCarthy had published *The Group,* a study in the banality of everyday life as pursued by a class of Vassar girls, circa 1933. It was her biggest and most ambitious novel, and although it became a huge best-seller, the critical reception was brutal: "She has failed from the center out, she failed out of vanity, the accumulated vanity of being overpraised through the years for too little and so being pleased with herself for too little," Norman Mailer intoned from the first page of *The New York Review of Books,* and then let loose the remarkable judgment that "she is simply not a good enough woman to write a major novel." McCarthy had swiftly become the second-most-vilified woman in American letters, not only artistically but morally. ("Combining being upset for you and upset for myself has made my head spin," she wrote to Arendt; "there is no cheek left to turn.") Sorely feeling her own lack of a public defender, she threw herself into producing a diatribe in support of Arendt. The article, titled "The Hue and Cry," appeared in *Partisan Review,* and was in some parts so morally devastating and in other parts so clumsy that, while Arendt's publisher in Munich requested its use in the German edition of *Eichmann,* everywhere else it only made her troubles worse.

"Hannah, let me tell you how I regret putting in Mozart and Handel," McCarthy appealed in June 1964, in response to the new wave of outrage her defense had prompted. In the most seized-upon of many provocations, McCarthy had compared *Eichmann in Jerusalem* to the uplifting final choruses of *The Marriage of Figaro* and *The Messiah,* calling the book "a paean of transcendence" in which "a pardon or redemption of some sort was taking place." What McCarthy had never imagined, as she burst out to Arendt, was "that anybody would use it to show that I was exulting over the murder of the Jews." McCarthy's choice of musical comparisons may seem as callow and perverse as her long-ago remark about Hitler. Yet, there is a passage in Arendt's exploration of the Holocaust to which the notion of redemption may be said to apply—an important passage that reverses Arendt's earlier declaration of the invincibility of the totalitarian state.

In the ultimate triumph of "radical evil," as Arendt had defined it in *The Origins of Totalitarianism,* individual acts of conscience or goodness disappeared into cunningly wrought "holes of oblivion," which destroyed not only individuals but the very meaning of their lives. By the time of *Eichmann,* Arendt was no longer concerned with the vast political origins of Nazism but with the small-scale human ones; as a result, she no longer saw evil as "radical" but as "banal"—that is, dull-spirited, occasionally even slipshod. In *Eichmann,* she does not refer to her earlier book or its conclusions. But in a late chapter she presents the single case of Sergeant Anton Schmidt, a German soldier who had helped Jewish partisans and then been executed. And she concludes that, after all, his actions did matter profoundly; that the resistance of even one man provides a moral and philosophical counterweight to the machinery of annihilation. Because we do remember Anton Schmidt, we do hold up his example. Because nothing made by men can function perfectly, and in this, at least, the Nazis labored in vain. A reader has the sense of enormous gears shifting direction when Arendt concludes: "The holes of oblivion do not exist." Experience has shown that there will always be someone who does not comply with terror, someone who lives to tell the story. If this hardly amounts to a chorus of redemption—if it seems an obvious truth to those who never doubted the innocent and who saw the greatest tragedy not in their corruption but in their slaughter—it is nonetheless a beam of light into Arendt's furiously dark view of humanity's future.

"About the Mozart business," Arendt replied to McCarthy's frantic apology, disclaiming the musical analogy but confessing that she'd loved it, "you were the only reader to understand what otherwise I have never admitted—namely that I wrote this book in a curious state of euphoria." Few, indeed, could understand what the diminished notion of evil meant to Arendt. Walter Laqueur wrote that she had been duped by the simple fact that all men "tend to be banal in prison." Gershom Scholem accused her of replacing a profound philosophy with a "catchword," and lamented what he called her "heartless" tone and her obvious lack of love for the Jewish people. Even McCarthy came to argue with Arendt's idea that Eichmann's evil had its root in "thoughtlessness"—by which Arendt meant a dependence on received opinion that prevented any real thinking from taking place. (Eichmann was, in Arendt's words, "genuinely incapable of uttering a single sentence that was not a cliché.") It was this insistence on the intellect's command of morality that McCarthy challenged, in a letter of modest wisdom that belies all the lurid dramatizing of her public defense. "I would have said that Eichmann was profoundly, egregiously stupid," she begins in agreement, but quickly goes on to offer her view that such "stupidity is caused, not by brain failure, but by a wicked heart. Insensitiveness, opacity, inability to make connections, often accompanied by low 'animal' cunning. One cannot help feeling that this mental oblivion is *chosen,* by the heart or the moral will."

There were many who believed—who still believe—that such mental oblivion was exactly what Arendt had chosen for herself. Her name remains a spur to argument in the dwindling pages of the newsletter of the American Gathering and Federation of Jewish Holocaust Survivors, and wherever else Jewish memory contradicts her account and calls itself betrayed. In defense of Arendt's intentions, it is clear from her earliest writings that, for her, even the awful burden of responsibility was preferable to the humiliation of the helpless victim. To deliver blame to one's people was to deliver control; if we allowed this to happen, then we can prevent it from happening again. And yet, Arendt's pitiless righteousness toward those whose choices she had never faced, and the tone that Scholem all too accurately described as "almost sneering and malicious," make reading her judgments, even today, an unnerving experience.

Throughout *Eichmann,* one is aware of the fact that Goethe had made Arendt into a German long before Hitler made her into a Jew, and the synthesis seems as unstable emotionally as it had been historically. Scholem informed Arendt that he considered her "wholly as a daughter of our people," while Jaspers proudly claimed her accomplishments—as he did Rembrandt's and Spinoza's—for Germany. Both were right, in a way, of course: as a highly cultured German Jew, Arendt was a being of near mythological conjoined endowments, merged at the soul and cruelly bred to be its own inquisitor; one of the last of a radical human experiment maintained for a few generations and still, years after Hitler, giving off a brilliant light fuelled by the energy of its self-consumption. "I have never in my life 'loved' any people or collective—neither the German people, nor the French, nor the American, nor the working class," Arendt vouched to Scholem. Certainly this was one way out of the difficulty: to renounce everybody equally. But she went on to specify the single group she renounced above all others. In a statement issued from what seems the psychic vortex of "the Jewish problem"—a statement that would have made Rahel Varnhagen wince in recognition—Arendt made it clear to Scholem that "this 'love of the Jews' " that he so absurdly demanded of her "would appear to me, since I am myself Jewish, as something rather suspect. I cannot love myself or anything which I know is part and parcel of my own person."

"Yesterday I read your exchange with Scholem in the new *Encounter.* I thought your part was very good, but I didn't like his tone of infinite sad wisdom," McCarthy wrote from her new home in Paris, just after Christmas, 1963. "I told George Weidenfeld, whom I saw for lunch the other day, that the net effect of all this controversy was to make me resolve never to set foot in the state of Israel. He was shocked." Modest wisdom clearly hadn't a chance when faced with a golden opportunity to do what McCarthy still liked to do best. And opportunities to shock were growing scarce since she'd settled into the role of the wife of a United States government official. In most ways McCarthy adapted exceedingly well to the settled life, fairly glowing with a ripe new matronhood—which she attributed to her first truly grown-up love for a grown-up man (compared to whom "Bowden is a child, and Edmund is an old woman"). Her newfound happiness was also clearly abetted by the tour-de-force dinner parties and luncheons and picnics that her wifely position afforded, and that seemed to bring her life to a kind of domestic apotheosis.

It soon appeared, however, that not even the pleasure of battle with imperious Parisian fruit vendors and surly Parisian maids could provide McCarthy with the regular test of moral will that she required: "I've been longing to get into a fight with someone," she informed Arendt the following Christmas. "France or Sartre and Simone. It is the same thing." But the real fight was unfolding back in America. In April 1965 McCarthy wrote to Arendt about her "doubt and dismay" over Vietnam, and worried whether there wasn't, in America, "any movement of protest among private people." (Awareness of the war was still so low that Arendt replied that she knew very little about it, and—perhaps through a linguistic error—placed Vietnam in Indonesia.) It does not take away from the meritoriousness of McCarthy's concern to note that she had long been waiting for just this kind of political situation—that is, for a chance to face up to a powerful force of evil, and to show what she was made of. Publicly, she hung back for a while, out of fear of compromising her husband's job. But then, in February 1967, she accepted an assignment from *The New York Review of Books* to report from Vietnam. By that time, there were four hundred thousand American troops on the ground. "I am leaving tomorrow morning," she began a particularly hurried, loving letter to Arendt. "I hope you approve. . . . I had to do *something*."

McCarthy considered the two books that resulted from her series of articles—*Vietnam,* set in and around Saigon, and *Hanoi*—as "pamphlets," in the most politically old-fashioned and potent sense of the word. The importance she attributed to this category of writing is clear when she states, at the end of *Vietnam,* that the war might be stopped by groups that included Johnson or Ho or a Republican president or "the readers of this pamphlet." And yet for all her declaration of urgent worldly purpose, these works seem hardly less personal than McCarthy's novels and essays—*Vietnam* opens with the words "I confess"—and the real target of her wrath seems not American foreign policy but American cultural squalor.

To read McCarthy, the war would seem to have been fought over differences in style. Her Saigon is a shoddy Americanized city, its cheap office buildings "teeming with teazed, puffed secretaries" and men "in sport shirts and drip-dry pants," its outlying villas rented to leathery middle-aged men in visor caps and their "cotton-haired" wives; damnation lurks in the lineup of ballpoints in a checked shirt pocket. The North Vietnamese, by contrast, are beautifully dressed, courtly of manner, graceful and young. They speak a superior French and easily quote both Terence—in Latin—and Noam Chomsky. The leader Pham Van Dong wears "a freshly ironed North Vietnamese army jacket of gray-tan poplin" and presents the intrepid lady reporter with a few well-chosen flowers from the palace gardens; on the streets outside, women streak by on bicycles in their pink or robin's-egg blue raincapes. This is a watercolor revolution, painted-fan Communism, as artfully smudged as the Trotskyist youth of the middle-aged Mme West, who arrived in Hanoi wearing a Chanel suit and bearing enough luggage to crush a caravan of bicycles.

It is hard to imagine that McCarthy would have ventured to Vietnam without the example of the uproar over *Eichmann,* which thrilled her with its evidence of journalism's possible impact as much as it upset her on behalf of her friend; the only impact her Vietnam books had, however, was to add the charge of political dilettantism to those already lodged against her. In fact, it is equally unlikely that she would have gone to Vietnam if her new novel had been going well. *Birds of America,* her ambitious attempt at a new-world *Candide,* was begun in 1964 and continually interrupted as McCarthy despaired of her ability to carry it off. Published at last in 1971, the book was dedicated to Arendt, and it is easy to see why. Apart from some sharp McCarthyan observations by the young half-Jewish hero—"Being an American," he reports while touring Europe, "was like being Jewish, only worse; you recognized 'your people' everywhere in their Great Diaspora and you were mortified by them and mortified by being mortified; you were drawn to them, sorry for them, amused by them, nauseated by them"—this is a novel of conscientiously big ideas, filled with the kind of ingenuous philosophizing and moral puzzlements that mark McCarthy's long correspondence with Arendt. Unfortunately, the ideas far outweigh the characters who speak them, and the result seems more an eager student's work than anything McCarthy had written in her early years. In the final scene, Kant himself makes a personal appearance at the hero's bedside—McCarthy had written Arendt about finding a portrait so she could describe him—and chats about the absent god and the end of the world in pidgin German.

The critical reception of *Birds of America* was predictably derisive, and this time there was no best-seller list to help make up for it. The book's commercial failure wasn't due so much to the insistent philosophizing (which never got in the way of Ayn Rand) but to the absence of

McCarthy's warmly vicious old humor and, above all, to the novel's complete lack of sex—a daring omission for a would-be best-seller, although it might be noted that throughout "late McCarthy" sex is almost entirely replaced by lingering, voluptuous descriptions of food. (Perhaps this had something to do with the influence of Paris, or it may have been simply a revenge for her early forced familiarity with the root vegetable.) Altogether, though, McCarthy seemed to believe that by omitting everything she had once been best at, she might be transformed at last into a novelist in the grand tradition; if only she were a good woman—did Mailer know he had hit the tenderest spot?—she might be a true successor to George Eliot. But McCarthy broke out of her narrow social confines only to discover that they were her talent's natural home. Her later efforts to attain moral gravity by covering the Medina trial—her clearest attempt to duplicate *Eichmann*—and then Watergate, only emphasized her inadequacies of scope and depth. Nothing in her letters to Arendt rings more unhappily true than the words with which, in 1975, the woman who corrected everyone and conquered everything confronted her limits: "It is sad to realize that one's fictions, i.e., one's 'creative' side, cannot learn anything. *I* have learned, I think, but they, or it, haven't. . . . It all leads to the awful recognition that one *is* one's life; God is not mocked."

"Times are lousy and we should be closer to each other," Arendt wrote McCarthy in February 1968, when she was finding it hard to hold on to the sense of moral shelter that America had once offered. Deeply discouraged now by the war in Vietnam, Arendt added that Johnson seemed "not just 'bad' or stupid but kind of insane"—a verdict she had never been willing to deliver with respect to Eichmann. Although she and her husband went so far as to consider moving to Switzerland, she refused to join in the resistance activities McCarthy urged upon her. ("I signed one of the many protests and that is that," she wrote. "I don't intend to make a profession of it.") Although her public writings continued to address a variety of political problems, and were winning her a university audience free of past arguments and animosities, in private she was increasingly concerned with a single issue: the fate of Israel. Arendt never relented in her criticism of the very foundations of the Jewish state, but in these years she reacted to each new threat to its safety with a fearful urgency. "One could not live there, of course," she wrote McCarthy, and yet "I know that any real catastrophe in Israel would affect me more deeply than almost anything else."

When her husband died in October 1970, Arendt barely kept herself from giving him a Jewish funeral, out of her longing to hear the Kaddish. (To McCarthy she explained the meaning of the Jewish death prayers, which praise God but do not mention the dead: "To be taken away it had first to be given. If you believed you owned, if you forgot that it was given, that is just too bad for you.") McCarthy arrived from Paris immediately to be with her. A few days later, a poem of affectionate condolence—entitled "Time"—arrived from Martin Heidegger. The Blüchers had visited the Heideggers just the year before, in response to Heidegger's renewal of communications in the autumn of 1966. (It is perhaps relevant, if not cynical, to note that in February 1966 *Der Spiegel* had published an article critical of Heidegger's Nazi past and his anti-Semitism. "I didn't like it at all," Arendt wrote at once to Jaspers; "He should be left in peace.") The visit of the two couples had gone well and the Blüchers had been planning to return. Instead, Arendt went alone to see the Heideggers the following summer, and again every year after that for the rest of her life. To mark Heidegger's eightieth birthday, she published an essay that worshipfully exonerated "the hidden king" of the realm of the mind, and recalled that Plato had also offered his services to a tyrant.

There seemed to Arendt no contradiction between the intensity of her belief in Heidegger's work and her renewed feeling for the Kaddish, or between what these oppositions represented. Both were ways back home. She was overjoyed when, in 1975, a few months before her death, she was invited to celebrate Passover among friends who had not welcomed her since the publication of *Eichmann*. Arendt's last years have something of the pathos of the prodigal, and it is hard not to think of her own "Rahel Varnhagen," which begins with her heroine's deathbed cry: "What a history! . . . The thing which all my life seemed to me the greatest shame, which was the misery and misfortune of my life—having been born a Jewess—this I should on no account now wish to have missed." It is clear only now that Varnhagen's full works have been published that Arendt actually altered the quotation by a simple act of omission, choosing to leave out her heroine's final emotional reversal and reaffirmation of a life devoted to the saving grace of Jesus Christ.

When Arendt died, the piece of paper rolled into her typewriter was headed "Judging," and its only text was a pair of epigraphs, from Lucan and from Goethe. The task of assembling this essay from lecture notes and then editing and "Englishing" its companions—to produce Arendt's final volume, *The Life of the Mind*—fell to Arendt's literary executor, Mary

McCarthy. Putting aside her latest novel and her ever-deferred dream of a study of the sculpture of Gothic cathedrals, McCarthy spent two years of mixed drudgery and elation on the project. Without the letters, one would not know that McCarthy had read much of the material before and had disagreed sharply over a central point: the human faculty in which a sense of morality is born and resides. Arendt had argued for the mind, McCarthy for the heart, each finally trying to locate something not so very different: the elusive source of imaginative sympathy by which we think or feel ourselves into the flesh of another. But each woman, unyielding, continued to hold up her holiest emblem, like the figures of Synagoga and Ecclesia on the Rhenish churches that McCarthy never managed to write about, except to Arendt. "I find the Synagogue figure very poetic," she had written affectionately just a few years before Arendt died. "She is almost always more beautiful in her melancholy than the Church, her sister." Lacking McCarthy's study of these figures, which would doubtless have evoked a more finished portrait of the friendship, these letters must do instead. As the correspondence closes, there is some consolation in the larger, architectural view: across the portal of literature, before the cathedral of history, the two sisters straining upward and holding forth in loving disagreement until the Day of Judgment.

Index

Illustration Credits

Olive Schreiner: Durham University Library

Gertrude Stein: Photograph by Carl Van Vechten, reproduced by permission of the Estate. Denison Library, Scripps College

Anaïs Nin: Photograph by Carl Van Vechten, reproduced by permission of the Estate. The Beinecke Rare Book and Manuscript Library, Yale University

Mae West: Photofest

Margaret Mitchell: Courtesy of the Atlanta History Center

Zora Neale Hurston: Photograph by Carl Van Vechten, reproduced by permission of the Estate. The Beinecke Rare Book and Manuscript Library, Yale University

Eudora Welty: The Granger Collection, New York

Marina Tsvetaeva: Courtesy of Ardis Publishers

Ayn Rand: Copyright © Arnold Newman

Doris Lessing: Copyright © I. Kar/Camera Press/Retna Ltd. USA

Hannah Arendt and Mary McCarthy: The Hannah Arendt Literary Trust

A Note About the Author

Claudia Roth Pierpont, a contributor to *The New Yorker* since 1990, has been the recipient of a Whiting Writer's Award and a Guggenheim Fellowship. She holds a Ph.D. in Italian Renaissance art history from New York University.

A Note on the Type

This book was set in a version of Monotype Baskerville, the antecedent of which was a typeface designed by John Baskerville (1706–1775). Baskerville's types, which are distinctive and elegant in design, were a forerunner of what we know today as the "modern" group of typefaces.

Composed by Dix,
Syracuse, New York

Printed and bound by Quebecor Printing,
Martinsburg, West Virginia

Designed by Cassandra J. Pappas